YA

Dreaming in
CODE

Dreaming
in
CODE

ADA BYRON LOVELACE, COMPUTER PIONEER

EMILY ARNOLD McCULLY

CANDLEWICK PRESS

The author would like to thank Charlie Carter,
at the Pforzheimer Collection, New York Public Library; copyeditors
Susan VanHecke and Hannah Mahoney; designer Lisa Rudden;
and above all, Candlewick editors Mary Lee Donovan and Kharissia Pettus.

First edition 2019

Library of Congress Catalog Card Number 2018961266
ISBN 978-0-7636-9356-5

18 19 20 21 22 23 LSC 10 9 8 7 6 5 4 3 2 1

Printed in Crawfordsville, IN, U.S.A.

This book was typeset in Crimson Text.

Candlewick Press
99 Dover Street
Somerville, Massachusetts 02144

visit us at www.candlewick.com

For Liz

CONTENTS

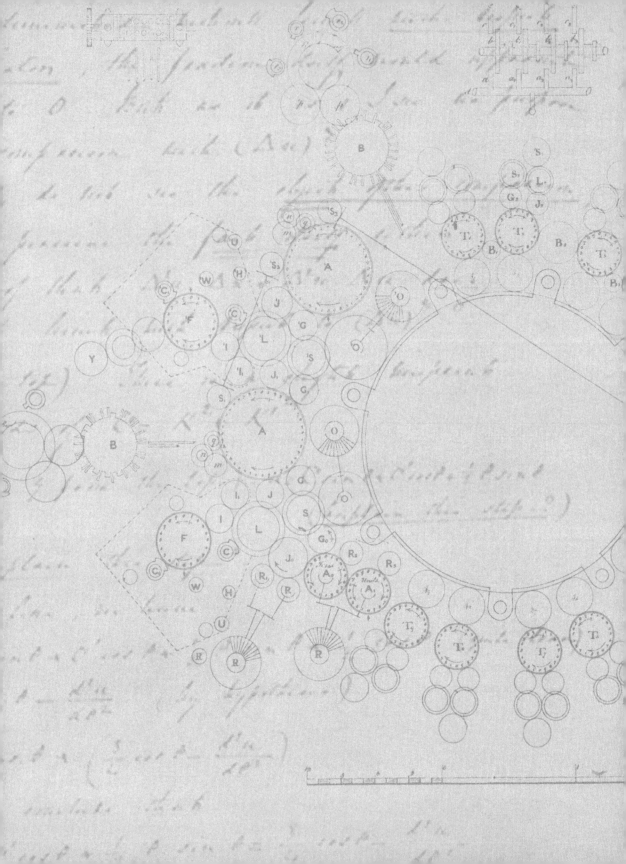

INTRODUCTION

Ada Byron Lovelace has only recently been receiving the recognition she earned nearly two hundred years ago. Brilliant, mercurial, and ambitious, she was born in an era that considered women's minds unequipped for serious thought. That wasn't the only barrier she had to overcome. Almost from birth, her mother set out to mold and then to control her. Ada seized her independence but was tormented by chronic ill health.

Yet she thought she possessed singular qualities of genius and aspired to outshine her famous poet father, Lord Byron. She knew she was an intuitive "discoverer of the *hidden realities* of nature," with "immense reasoning faculties" driving her toward an uncharted destiny.

She and the mathematician and inventor Charles Babbage formed a collaboration that was a kind of alchemy. Her energy, boldness, "Fairy" magic — he called her an "Enchantress" — pushed him to

A portrait of Ada Lovelace, age twenty

develop his "Analytical Engine," an early mechanical computer. Then she explained the invention to the world. But the world wasn't ready to imagine what Ada imagined — computers that would process information by themselves.

She seems to have realized, as Babbage did, that their work would be appreciated only by future generations. She wrote him:

> *You know I am by nature a bit of a philosopher, & a very great speculator, — so that I look on through a very immeasurable vista, and though I see nothing but vague & cloudy uncertainty in the foreground of our being, yet I fancy I discern a very bright light a good way further on, and this makes me care much less about the cloudiness & indistinctness which is near. — Am I too imaginative for you? I think not.*

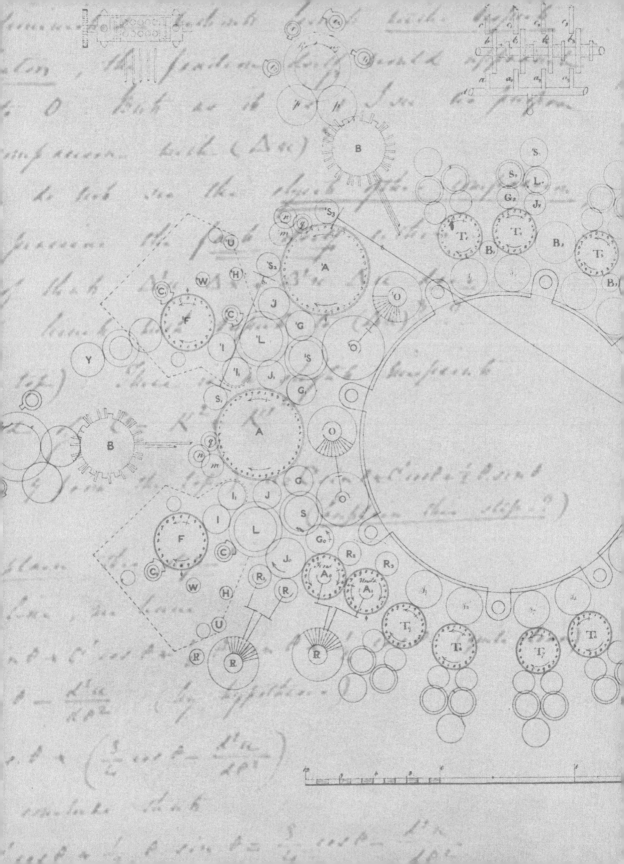

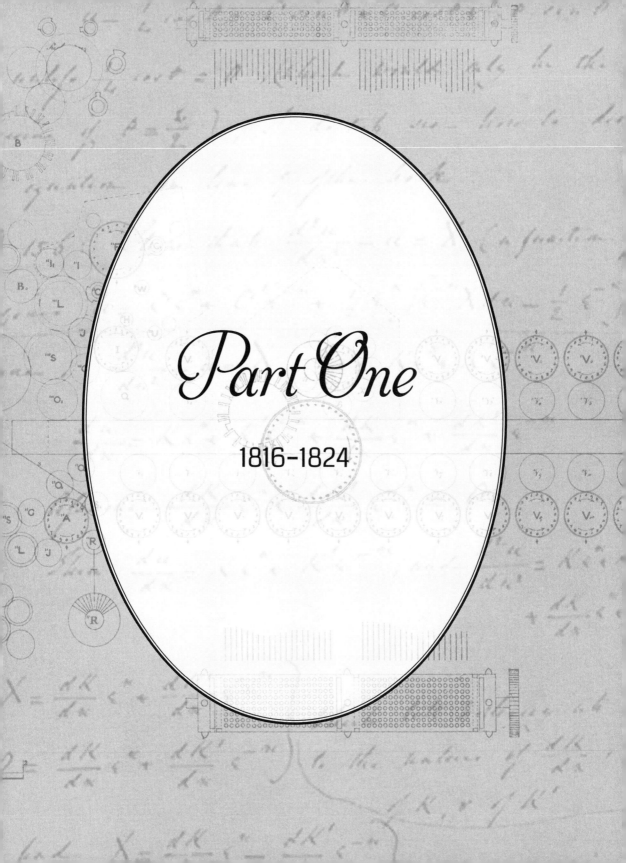

Part One

1816–1824

Chapter 1

Born into Scandal

Augusta Ada Byron was only a month old on January 15, 1816, when her mother, Anne Isabella Byron, snatched her from her cradle and fled London.

Lady Byron and her husband, George Gordon, Lord Byron, had been married just a year. During their months together, he was rarely tender — mocking her, howling that marriage had destroyed him, proclaiming his love for his half sister, confessing his past misdeeds, and recounting the madness that ran in his family. Often, he drank until he passed out.

Lord Byron was drunk the night Lady Byron told him she was leaving him in the morning and would take Ada with her.

He languidly asked, "When shall we three meet again?"

She answered, "In Heaven," and ran from the room.

* * *

Titled, handsome, reckless, and irresistible, Lord Byron was Europe's first pop superstar. He had become famous overnight with the publication of *Childe Harold's Pilgrimage*, a bold novel in verse about the travels of a hero resembling himself who searches for meaning in nature as the world of men disappoints him.

> *There is a pleasure in the pathless woods,*
> *There is a rapture on the lonely shore,*
> *There is society where none intrudes,*
> *By the deep Sea, and music in its roar:*
> *I love not Man the less, but Nature more,*
> *From these our interviews, in which I steal*
> *From all I may be, or have been before,*
> *To mingle with the Universe, and feel*
> *What I can ne'er express, yet cannot all conceal.*

The poem's beauty and romantic intensity, combined with its hints of autobiography, created a cult of Byron. Men practiced his scowl before their mirrors. Women threw themselves at him, and he responded with gusto. One lover called him "mad, bad, and dangerous to know."

Byron's future wife, Anne Isabella Noel, called Annabella, was an only child born to older parents who cherished her intelligence and had her tutored from the age of fourteen in literature, philosophy, astronomy, and especially mathematics. Annabella was also known to be a prim, religious woman with strict morals. So it struck others as odd that she married Lord Byron. She had known he was wild;

everyone did. A sharp-tongued *grande dame* took note of their engagement with this remark: "How wonderful of that sensible, cautious Prig of a girl to venture upon such a Heap of Poems, Crimes & Rivals." Annabella, however, was not really behaving out of character; she believed she could tame Lord Byron if they married.

Lord Byron did seem to be taken by the young woman's intelligence (if not by her priggishness), and he also seemed taken by her family's wealth. She was in line to inherit a fortune, and Lord Byron hoped it would restore his finances. He was a spendthrift, like his father before him. His careless extravagance meant he was chronically in debt. He and Annabella were married in January 1815. The inheritance was delayed, and there was no telling when it might come to Lady Byron.

The newlyweds had both made an awful mistake.

Lady Byron left, as she had said she would, in the morning. It was now up to her alone to raise the child of her bitter union.

Annabella had written out a perceptive analysis of Byron, noting his "habitual passion for excitement, which is always found in ardent temperaments." She added, "an Ennui of a monotonous existence . . . makes [him] seem to act from bad motives when in fact [he is] only flying from internal suffering. . . . Drinking, Gaming &c. are all of the same origin." (She noted later that her reflections about his character were "written under a delusive feeling in its favor.") Should hints of these manias appear in Ada, Lady Byron was determined to suppress them. Ada would not even be told who her father was; the child could not be allowed to form any attachment to him.

It wasn't easy for a woman to obtain a separation during that era, and custody of a child usually went to the father in a separation or divorce.

But rumors had already been circulating through London society that Byron engaged in homosexual acts (then a capital crime) and that he had fathered a child by his half sister. These allegations ensured that custody of Ada would almost certainly go to Lady Byron.

Lady Byron's lawyer arranged for Ada to be given over to her mother but as a ward of the chancery, the ancient court that oversaw estates. Lord Byron was insulted by this stipulation because it implied that he would try to take forcible possession of his child. He said he had no intention of doing so, but Lady Byron cultivated her role as victim and martyr by insisting that he did. Her great willpower and her desire to be viewed as a paragon of virtue made her a formidable figure, even

Anne Isabella Noel, Lady Byron, shown here as a prim young woman

in her early twenties. She was fashioning a career out of having suffered at Lord Byron's hands, while at the same time clinging to the aura of his fame.

Newspapers printed more stories of the "Byron Separation" than of any celebrity event in British history. It was the talk of Europe, and made the infant Ada an object of fascination that would last for the rest of her life.

Upon arrival at her parents' country estate at Kirkby Mallory, Lady Byron and Ada were welcomed with agitation and excitement, according to a resident of the Kirkby parsonage, where they had been following the drama in the press.

George Gordon, Lord Byron, was an English Romantic poet who enjoyed cultivating his reputation for wit, recklessness, and cruelty.

What people likely didn't know at the time was that, in the carriage on the way there, her tiny daughter asleep at her side, Lady Byron admitted to herself that she was not a natural mother and had no patience for infant care.

Chapter 2

Mother and Child
with Governesses

Despite her admission of maternal limitations, Lady Byron was loath to delegate Ada's care entirely to others. Yet whenever the nanny left her alone with Lady Byron, the child began to shriek. This may have been the result of Annabella's lancing her gums. "Ada has cut two more teeth with my assistance . . . and the Nurse is shocked at my Want of Mercy," she wrote, with want of mercy. Lady Byron was irked in turn by nurse Grimes's "selfish way of assuming an authority directly opposite to my wishes, and by the cruel manner in which she has treated every attempt of mine to regulate the Child's temper." (She always referred to Ada as "the Child.")

When Ada caught chicken pox, it was Nanny Grimes who stayed up night after night to nurse the child.

Ada naturally grew quite attached to her nanny, much to Lady Byron's irritation. Refusing to tolerate anyone else's strong influence on Ada, she finally had Grimes dismissed.

Ada adapted to her mother's behavior and eventually got over screaming whenever she appeared. The little girl then began looking for ways to win Lady Byron's approval instead. By the time she was two, she was mimicking her mother. The little parrot absorbed the lesson that the sure way to avoid offending Lady Byron was to copy her exactly. Lady Byron proudly reported this behavior to friends. When chastised, Ada sometimes flung herself on the floor, cried for a few minutes, then sprang up, saying, "Wipe away her tears . . . I good." Her exuberance made her charmingly impetuous. When first shown the sea, she exclaimed, "Throw me in!"

As soon as Ada could read, she was assigned *Duties of Children to Their Parents.*

For a while, Lord Byron hoped for a reconciliation and sent letters asking for news of his baby. Once he enclosed a poem including the poignant lines

When our child's first accents flow,
Wilt thou teach her to say 'Father!'
Though his care she must forego?

He confided in his publisher: "I have a great love for little Ada, and I look forward to her as the pillar of my old age, should I ever reach that . . . which I hope not."

In November 1816, just one month shy of Ada's first birthday, the third canto of *Childe Harold's Pilgrimage,* his serially published masterpiece, appeared. In it, he publicly expressed his longing to see his child with the lines

Is thy face like thy mother's, my fair child!
Ada! sole daughter of my house and heart?
When last I saw thy young blue eyes they smiled,
And then we parted,— not as now we part,
But with a hope.—
   Awaking with a start,
The waters heave around me; and on high
The winds lift their voices: I depart,
Whither I know not.

The poem was composed aboard a ship. The relentless gossip had driven Lord Byron from England. "My name has been completely blasted as if it were branded on my forehead," he wrote Annabella. He left London for Dover in a gilded coach copied after Napoleon's, pursued by agents of the many people to whom he owed money, including the coach maker, and sailed across the English Channel to Ostend, Belgium.

There was no hope of reconciliation. Lady Byron was adamant about excluding Lord Byron from their lives. A green cloth curtain hung over the mantel in the Kirkby drawing room. Ada was told never, ever to peek behind it, and servants hovered to enforce the rule. They knew what she did not — the green curtain concealed a portrait of Lord Byron. Lady Noel had sold some of her jewels to buy it in 1815.

Lady Byron threw herself into the role of separated and wronged wife. Henceforth, she would perform charitable acts that displayed her religious zeal, and she would instill morals and reason in her child.

The hidden portrait of Lord Byron in Albanian dress.
Ada was forbidden to look at it until she was twenty-one.

She resolved that her child's upbringing should be governed by a "rational order of education from the cradle" and drew up a program that would keep Ada's mind busy all day and free herself to pursue her interests. Lady Byron's own education had made her an intellectually confident young woman who knew she could supervise Ada's training in certain subjects. To her old tutor, Dr. William Frend, she wrote, "My daughter is a happy and intelligent child, just beginning to learn her letters — I have given her this occupation, not so much for the sake of early acquirement, as to fix her attention, which from the activity of her imagination is rather difficult."

When Ada was five, her mother wrote, "I have engaged a person to teach the elementary part of a few things which can now be commenced, during the next 3 months — as it is my object to avoid that permanent evil called a Governess — & by means of occasional assistance, and my own exertions, I hope to educate Ada sufficiently."

On the recommendation of a professional educator, she hired a tutor, Miss Lamont, who kept a daily log of Ada's progress in the schoolroom. The lessons began on May 14, 1821.

Lady Byron left detailed instructions: "Be most careful always to speak the truth to her. . . . Take care not to tell her any nonsensical stories that will put fancies into her head." Furthermore, there was a rigid daily lesson schedule: music, 10; French conversation, 10:45; arithmetic, 11:30; needlework, 1:30; music, 2:45; French exercises, 4:30. Ada was also taught to sew (and years later made her own petticoats).

Ada could already add complicated sums and read well, knew rudimentary geography, and could draw parallel lines and explain the terms *perpendicular* and *horizontal.* Her memory was unusually acute. Miss Lamont was clearly startled by her charge's well-developed sense of right and wrong: "At night she cannot go to rest without having examined herself, whether through the day she has committed any wrong act! — Her sensibility is easily worked upon." Still, little Ada was "brim full of life, spirit, and animation and is most completely happy."

Ada was not to leave the schoolroom until the governess had come in to grade her exercises. Even when she started on geometry, which she adored, it was hard to sit still, to not toy with some object at hand, to not tease her cat, Mrs. Puff. To combat Ada's chronic fidgeting, Lady Byron set up a system of rewards and punishments to keep her on track. If Ada completed her lessons and they were correct, she got a ticket of excellence. A dozen tickets could be exchanged for a book she wanted, sometimes even a novel. But if her work wasn't satisfactory, Miss Lamont took the tickets back and she had to start over again. When Ada drummed her fingers, bags were tied over her hands. If she squirmed, she sat in the corner for an afternoon, lay on her back on a board meant to improve posture, or, worst of all, was put in a dark closet.

On Sundays, Ada was allowed to make things out of wooden blocks by following the models of the blocks' designer, who was a Swiss educator and philanthropist Lady Byron admired. Blocks were an unconventional toy for a girl during that era, but Lady Byron was no ordinary mother. The appeal of the Swiss blocks lay in their having a program for their use. Ada's inclination was to be more original. When lessons were over, she constructed fanciful towers and bridges instead of the plain boxes prescribed and modeled. Lady Byron allowed it.

At pains to please her mother, whenever Ada's mind wandered or she made mistakes, she begged forgiveness. Once she drew up a table of her lessons with two columns, labeled "G" for good and "B" for bad.

In one of her obligatory letters to her mother, who was again away, she wrote penitently that she deeply regretted her bad work and promised to improve her character.

> *I want to please Mamma very much, that she & I may be happy together. . . . I was rather foolish in saying that I did not like arithmetic & to learn figures, when I did — I was not thinking quite what I was about. The sums can be done better, if I tried, than they are. The lying down might be done better, & I might lay quite still & never move.*

Despite Miss Lamont's best efforts, she was sent packing after a few months. As Lady Byron explained to the educator who had recommended her, "Miss Lamont appears quite unable to gain the necessary ascendancy over so masterly a mind as Ada's. . . . [She] seems quite overpowered . . . in circumstances which demand prompt decision." She ended chillingly, "The only motive to be inculcated with a character like Ada's is a sense of duty, combined with the hope of approbation from those she loves."

The experience had clarified for Lady Byron the conditions that she believed were necessary for Ada's education to succeed.

The two were sometimes locked in combat, but were still joined together in the most primary of human relationships, that of mother and child. Lady Byron believed she was raising her child in a superior manner, and in many ways, she was. Ada was getting a far more rigorous education than other girls at that time. On the other hand, Ada neither received nor learned much emotional nurturing.

Ada could count only on her old grandmother, Judith, Lady Noel, for kisses, hugs, and kindness. Lady Noel and her husband, Sir Ralph Milbanke, 6th Baronet and a member of Parliament, were opponents of slavery and supporters of relief for the poor. Lady Noel was a little afraid of her daughter, so refrained from challenging Lady Byron's stringent child-raising policies, compensating for them instead with expressions of affection. She told Lady Byron, "the *passion* She has taken for me is extreme, and she is always very good . . . and quite obedient to Gran-mama, who I assure you does not spoil her."

In another letter, Lady Noel gently encouraged her often-absent daughter to cultivate Ada's affections herself — Annabella composed a poem around that time, entitled "On a Mother Being Told She Is an Unnatural One." She wrote to a friend, "Ada loves me as well as I wish, and better than I expected, for I had a strange prepossession that she would never be fond of me."

Lady Noel died when Ada was six, a terrible loss for the little girl. It must have seemed almost unbearable when afterward Lady Byron told Ada they were leaving Kirkby Hall, the only home Ada had known. They went to live in the first of a series of grim, rented country houses with little in them that was familiar, inviting, or homelike. Lady Byron took no interest in decoration, gardens, or fashion, judging all of it frivolous. In these unfamiliar surroundings, Ada was forbidden to speak to estate workers or villagers, lest one of them be an agent of Lord Byron.

Chapter 3

She Has a Father

At age seven, Ada began to suffer terrible headaches, severe enough to interfere with her vision. Her mother, a confirmed hypochondriac, had a squadron of doctors on call and one of them diagnosed Ada with a "fullness of the vessels of the head" or possibly migraines. The doctor concluded that her rigorous lessons were putting too great a strain on her. Until she recovered, she was to lie all day in the dark. Lady Byron was also instructed to apply leeches to Ada's head, and she sometimes left them in place for nearly a week. It was a common cure, and one Lady Byron used on herself with particular relish. Ada endured it. In time, she got better.

It was Lady Byron who wished to be seen as chronically ill. Time after time, she told Ada she might die. Sometimes her cures at various spas lasted for months. During these absences, Ada was made to kiss her mother's portrait every day in the presence of a witness. It is little wonder then that Ada clung to her most constant companion: her cat, Mrs. Puff.

Although she ran afoul of Lady Byron with some regularity, Ada's "prevailing characteristic" managed to be "Cheerfulness — a disposition to enjoyment." She was a resilient little girl who was quick to learn and enjoyed it immensely. When rebuked, she bounced right back. Ada's intelligent, vivacious, curious, eager, and exploring nature was unimpaired. Her large, dancing eyes reminded people of her father's (although they knew not to tell her so), and her dark hair curled becomingly around her face. She was versatile, too, talented at drawing, music, languages, logic, and mechanics.

Lord Byron remained curious about his only child born of a marriage. In 1823, he wrote to his half sister Augusta Maria Leigh, from Greece, where he had gone to fight with Greek people in their quest for independence from the Ottoman Empire, to ask that she obtain from Lady Byron some "account of Ada's disposition — habit, studies . . . temper . . . personal appearance. . . . Is the girl imaginative? . . . Is she social or solitary? . . . Is she passionate? I hope that the Gods have made her any thing save *poetical* — it is enough to have one such fool in a family."

Ada was also unaware that Lord Byron had repeatedly begged Lady Byron, through intermediaries, for a portrait of her. In 1824, Annabella produced a detailed description of Ada and sent it along with a silhouette. Byron proudly showed it off to his companions at a garrison in Greece.

A miniature portrait of Ada, age five, during her schoolroom years

The following April, he was dead, not by a battle wound but of illness aggravated by bloodletting and unsterile medical instruments.

Masses of people in Europe and America plunged into deep mourning. Byron had been considered the greatest poet alive. Eminent writers, when asked to comment, were too distressed to pick up their pens. In England, the shock and grief was felt in every part of society. One would have to have been living in a cave not to be aware of his art and notorious escapades and feel the enormity of his loss.

His own daughter, however, had never even heard his name. She knew that other children had fathers and that she did not, but why not? She did have a grandfather, so one day when she was seven or eight, she had blurted out what was for her a logical question, asking: "if Grandpapa & Papa were the same." Lady Byron delivered a scolding so severe that Ada was numbed by it. She was left with "a feeling of dread toward her mother that continued till the day of her death," according to her friend Woronzow Greig.

When Lady Byron told Ada that her father was dead, Ada was doubly stunned: The father she had longed for was real but already dead? The hard shell enclosing her childhood was finally cracked. Fixing her eyes, so like his, on her mother, Ada began to weep. Lady Byron, self-centered as ever, reported to a friend that Ada must surely have wept out of sympathy for *her*. After all, how could a child cry for a father she hadn't known?

Lord Byron's body was returned to London, where the news of his death had arrived earlier "like an earthquake," as the press put it. While his body lay in state, thousands of Londoners came to pay their respects. Meanwhile, a debate raged over whether or not to inter him in Westminster Abbey in the corner reserved for poets. Church elders vetoed that. It was decided to bury him at his family's residence.

When the hearse and forty-seven carriages departed London for Newstead Abbey, the Byron estate in Nottingham, mourners lined the streets. Crowds of local residents wearing mourning dress greeted the hearse when it arrived at five in the morning and followed its progress along flower-strewn roads all the way to the church funeral at Hucknall.

Ada was told only that her father had been an eminent poet yet a wicked man. It would be years before she saw the verses Byron had written about her:

> *Ada! sole daughter of my house and heart?*
> *When last I saw thy young blue eyes they smiled.*

Knowing at last that she had had a father seemed to change Ada. She began to slip her bonds, little by little, and come into her own in thought and feeling. She challenged her mother's unyielding rectitude, pointing out the flawed logic of Lady Byron's incentive system and arguing very sensibly that virtue should be its own reward.

"I should wish that . . . you do not give me a reward because I think the reward of your being pleased with me sufficient besides when you do that I don't do the good thing because I know I ought to do it but because I want to obtain the reward, and not because I know it to be right, and if I was encouraged in this, when I was grown up . . . I should never do any good without I had a reward."

This was an early appearance of what Lady Byron called Ada's "conversational litigation." (Over her long life, Lady Byron herself was quick to file a lawsuit against anyone she thought had crossed her. And she disliked being contradicted by her daughter.)

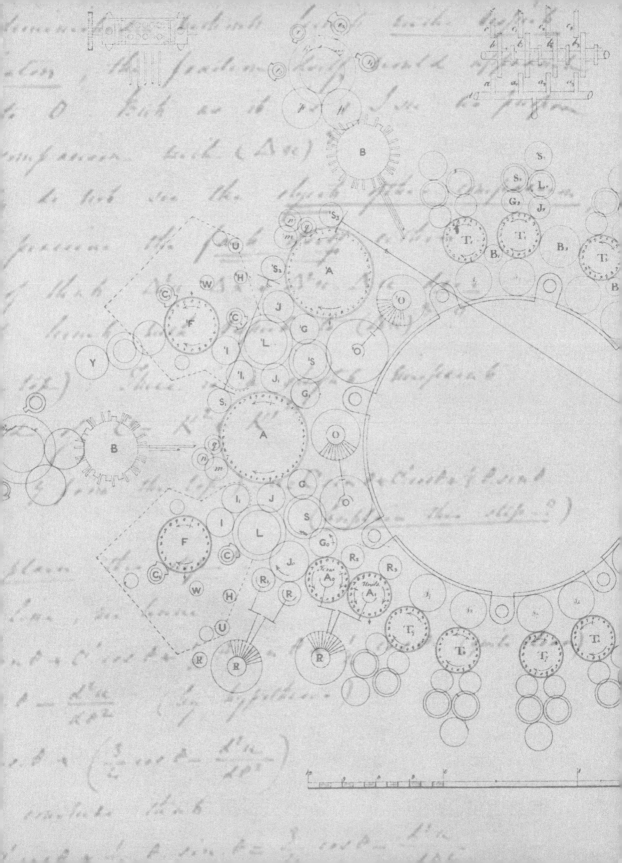

Part Two

1825–1832

Chapter 4

Her Imagination Soars

Molding and educating Ada was Lady Byron's main focus until 1825, when, a year after Lord Byron's death, she finally inherited her family's enormous fortune. Her new holdings included coal mines in the north of England. Coal powered the new railroads and produced the steam that made mass production of goods possible in factories. With these means, she began redirecting some of her energies from Ada's edification to ambitious philanthropy.

Coal was dug by miners, many younger than Ada, who labored punishingly long hours underground, breathing thin, poisonous air and dying of lung diseases and malnutrition, equipment accidents, or explosions. Industrialization destroyed businesses in which goods were handmade and sent people from farms to cities, where they sought factory jobs that undermined family life and health.

Lady Byron had been raised by her parents to be concerned about the world and so joined the growing band of church-affiliated

Lady Byron's depiction here suggests her self-view as long-suffering and magnanimous. She began an ambitious philanthropy campaign after she inherited her family's fortune.

reformers determined to help the disadvantaged, including the workers connected to her new properties, and to mitigate some of the problems resulting from the Industrial Revolution. She directed a good portion of her wealth to founding and overseeing schools for the children of poor laborers and farmers. Applying the latest in education theory, her schools included work projects for the children and even markets for the goods they produced.

While her mother was off managing her charities, Ada dutifully wrote letters to her, reporting that she was studying French and Italian, writing a description of arrowroot, and working out the solution for the problem "If 750 men are allowed 22500 rations of bread per month how many rations will a garrison of 1200 men require?" She also tackled decimals with her current governess, Miss Stamp.

Ada's imagination was as active as ever, and her mother didn't always discourage it. Clearly inspired by Lady Byron's charitable enterprises, Ada described a plan for a utopian colony for orphaned children. She promised, as she often did, to reform her own character and to cultivate a compassionate understanding of others so that she'd be properly qualified to run her colony. To her letters, Ada would add that she looked forward to her mother's return and then ask if she might wake her up earlier in the morning so that they could have more time together.

Ada spent most of her life trying to please her unbending judge. Lady Byron's intelligence made the effort worthwhile, however disheartening it could be at times. There is great satisfaction in winning over a smart critic, and when Lady Byron came around, it was a triumph for Ada.

Ada's mercurial yet resilient temperament, her exuberance and enthusiasm, her curiosity and precociousness, all kept her from being entirely intimidated by her mother. She was increasingly aware of her own strangeness, the quality that later made her a "Fairy" and an "Enchantress." Ada was an exceptional little girl, driven by deep urges and insistent voices.

In 1827, when Ada was eleven, she and her mother went on a European tour. A few of Lady Byron's friends and Ada's governess, Miss Stamp, went along.

Ada loved being abroad; every new setting delighted her. She was especially struck by the steamships on Lake Geneva in Switzerland and by the play of colors on the water's surface, which she studied for hours. She made chalk sketches of dramatic Swiss landscapes. The organ music in great cathedrals thrilled her. She was freed of the usual lessons, but required to write regularly to her mother's friends in England, and those letters had always to include a report on Lady Byron's health.

Even abroad, Ada continued to fascinate as the child of Lord Byron and was often stared at. To make matters worse, around this same time, one of Lady Byron's friends told Ada that she wasn't pretty. These things combined to make her uncomfortable and painfully self-conscious. As she often did when faced with discouragement, however, Ada decided to draw a positive lesson. She reasoned that vanity was the cause of most unhappiness; not being pretty therefore meant she was spared both vanity and unhappiness.

After more than a year of travel, they returned to England,

where Lady Byron rented a house called Bifrons, outside the city of Canterbury. She didn't stay there long herself: a vague, disabling lassitude made her seek another cure at a spa, this one lasting several months. She left Ada at Bifrons with Miss Stamp.

There, a new enthusiasm seized Ada: *flight,* the very essence of freedom.

Hot-air ballooning, which had started in France at the end of the previous century, was now popular in England. One balloon had just famously crashed near London. But Ada's vision was not to sit passively in a basket and drift about among the clouds. As a child of the Industrial Revolution and product of a regimented course of rigorous study, it was also her nature to think both imaginatively and scientifically, working through problems by conducting experiments when possible. Ada observed the creatures who *could* fly, and, like Leonardo da Vinci before her, analyzed the structure of a bird's wing. Then she tried to build a pair of wings, predicting that it would take a year of patient experience and practice to "bring the art of flying to very great perfection."

Her wings would be made not of paper but of oiled silk. If silk made them too heavy, she'd try feathers. She was working out how to attach the finished wings to her shoulders and asked her mother to send her a book on bird anatomy. Once Ada was airborne, she'd have to be able to steer, but she had an idea for that already. If her methods worked, she would write a book and call it *Flyology.* Ada reported every step of her process in letters to her mother, signing herself "Carrier Pigeon."

Ada also wrote that she was thinking of practical applications for flying — mail delivery was one. To her design, Ada had added steam power, the phenomenon that was mechanizing industry and travel. She would cast a metal horse body and install a steam engine inside it. The craft would be ridden as one would ride an actual horse. She acknowledged that there were obstacles to its succeeding but cheerfully noted that the thought of riding a steam-powered horse through the sky had encouraged her to ride her own horse more often. Lady Byron was bound to approve of that. Young ladies of rank had to be proficient riders.

Although Ada was still isolated from nearly all that was happening outside her tiny sphere, she had caught the spirit of the age, which was daring, inventive, and dazzled by the potential of technology.

Lady Byron was fairly patient with this obsession, but in time she told Ada that she had wasted far too much time on flying and put an end to her flyology.

Chapter 5

Guarded by Furies

Never content to stay in one place long, Lady Byron leased a new villa called Hanger Hill, in the Ealing district of the London suburbs. At the same time, Ada's governess, Miss Stamp, resigned because she was getting married.

Instead of replacing Miss Stamp with another governess, Lady Byron hired a team of educators to supervise Ada's studies. One was elderly Dr. William Frend (who had been Lady Byron's own math tutor decades before). He brought Sophia, his daughter, to help monitor Ada's behavior. The third was a devout physician, Dr. William King.

"There are no weeds in her mind," Lady Byron wrote to King about Ada. "It has to be planted. Her greatest defect is want of order, which mathematics will remedy. She has taught herself part of . . . *Geometry*, which she liked particularly." Two other tutors were recruited from the industrial schools movement. Like Frend and King, they were religious and conservative in their views.

In 1829, before her schooling could even begin, thirteen-year-old Ada caught measles and was put to bed.

The measles morphed into a more serious illness. Ada became so ill that she was unable to walk for nearly two and a half years. No one seems to know for sure what her affliction was, but the cure her mother ordered was prolonged bed rest, which itself would have severely weakened her muscles, even enough to impede walking.

In any case, it was no vacation for Ada. In fact, Lady Byron saw Ada's being bedridden as an opportunity to load on more studies. Miss Lawrence, mistress of a school in Liverpool, was added to the roster of instructors and worked with Ada mostly by mail.

William and Sophia Frend continued to tutor her in mathematics. Sophia later recalled that Ada "had *no* taste for poetry, but an exquisite appreciation of music, and afterwards became a good & scientific musician. She had as well great mathematical power. Her truthfulness was I think very questionable, where her vanity which was excessive was concerned, but she was exceedingly good-natured, and, in some ways kind feeling."

Lady Byron commonly sought the advice of deeply religious people, especially Dr. King, her personal physician and founder of the cooperative movement, wherein workers joined together to produce goods and share expenses as prices rose in the new capitalist economy. King was enlisted to give Ada moral guidance while she convalesced. He recommended geometry lessons to keep her mind occupied, and

Ada worked its theorems with relish. On her own, she taught herself German and read *Macbeth* as translated into German by Schiller even before she read the play in English.

Throughout her recovery, Ada was allowed only short stints outdoors in her wheelchair, and the frustrations of a long illness endured in adolescence began to tell. She was short-tempered and contrary in numerous small ways. One example was a refusal to sleep in her bed. She would lie on a sofa instead ("like a fish," she said) or on the floor wrapped in a blanket.

William King's authority as Ada's spiritual guide was absolute.

At sixteen, she was still recovering and using crutches when Lady Byron moved them yet again to Fordhook, another villa in Ealing. On the edge of the village green, this house was considerably more pleasing than the others had been. While Lady Byron went away to inspect reformatories, penitentiaries, hospitals, asylums, and even a convict ship, her circle of unmarried friends was always on hand to keep Ada in line. Their rigid rules were a goad, tempting her to test their limits. Traits that provoked her mother and her friends included picky eating habits, questionable truthfulness, and impertinence.

Ada called her mother's friends the "Furies." Later in life she told a friend that the Furies hated her like poison and told invented and exaggerated tales of misbehavior to her mother.

Chapter 6

An Elopement

In 1832, Ada was finally well enough to resume horseback riding, dancing, and gymnastics. She recovered her spirits too. Sophia Frend tartly observed that Ada "had the same love of startling & surprising people by her statements that her father had." Lady Byron and the Furies felt that her restless intellect had to be strictly channeled in order to prevent emotional outbursts. For that, there were lessons in chemistry, Latin, music, and, fatefully, shorthand.

England was in the midst of a shorthand craze. Shorthand is a fast way to write by using abbreviations and symbols. It was perfectly in keeping with the era's rage for speed and efficiency and the need to stay afloat in a flood of new information. Ada would use it to take notes from books and scientific journals or in lecture halls for most of her life.

So that her daughter could acquire this useful tool, Lady Byron hired a shorthand tutor named William Turner, an Ealing neighbor from a modest academic family. Little is known about him, but one thing can be inferred: he was near enough to Ada's age to be a potential romantic partner.

The rules of propriety in courtship were very strict at the time; Ada was to be matched to a suitor approved by Lady Byron. And then marriage was the only outcome. In addition, any romance between Ada and William would have jeopardized William's employment and his family's social standing.

Ada was watched constantly by one Fury or another. Nevertheless, she managed to slip him a note one day proposing that they meet later that night. Years afterward, Ada told a friend that matters "went as far as they possibly could without connexion." Passion made Ada and William careless. Someone witnessed an intimate gesture between them and reported it to Lady Byron. William was promptly dismissed. A disappointed and indignant Ada then ran away and into the arms of her lover, who had retreated to his family home. The Turners promptly returned Ada to Fordhook. Ada's behavior must have awakened her mother's worst fears — no behavior was more Byronic than seduction.

Not surprisingly, Lady Byron concluded that Ada's escapade proved the need for greater discipline. Dr. King, the religious advisor, was enlisted to lay out a program to curb high spirits and "objectionable thoughts." Much of the program consisted of sermons delivered on long walks.

In answer to her many questions about propositions and proofs, he offered more religious tracts. She also asked for books on optics and astronomy and told him she needed to master more math in order to understand these topics. (She needed "hard work.") Dr. King did his best to respond, but he had little expertise and told her, "You will soon puzzle me in your studies."

She was careful to assure Dr. King that mathematics was promoting good behavior: "I must cease to think of living for pleasure or self gratification," she wrote. "There is but one sort of excitement . . . which I think allowable for me . . . that of study and intellectual improvement. I find that nothing but very *close* & *intense* application to subjects of a scientific nature now seems at all to keep my imagination from running wild, or to stop up the void which seems to be left in my mind from a want of excitement." Ada's spirit wasn't broken. She challenged Lady Byron's assertion that she had been "constituted my guardian by God *forever*" and even warned her mother that she was obliged to obey her orders only until she was twenty-one years old. After that, unless her behavior interfered with her mother's comfort, she had the right to decide for herself what to do.

England itself was in a period of social upheaval in the mid-1800s. As industrialization accelerated, new cities swelled with people fleeing rural areas where there were no jobs. These new masses needed services but had no representation in the government.

Lady Byron ran her charities on the periphery of a much broader movement to create greater equality across rigid socioeconomic statuses, lest the lower ones rebel. A bill extending voting rights to more of the male population was repeatedly presented, along with measures to reform Parliament itself. In the time that the bill was debated and stalled, governments rose and fell, and frustrated workers rioted. Many feared the new urban poor would launch a revolution or that the rich would sabotage the economy by withholding taxes and draining the banks. There was radical talk of abolishing Parliament altogether and even of removing the monarchy.

Lady Byron had no concerns about her own social standing. She described herself as handling crises "calmly." "I trust to the principles on which I have always acted towards my inferiors. . . . I have not learned to like riches for their own sake — and I *dare* be poor." She told Dr. King she hoped there would be a redistribution of property and was willing to give up some of hers, but only to people who had the right attitude. "I would rather they should take 3/4 of my property in the *right* spirit than 1/4 in the *wrong*."

In the spring of 1832, the Great Reform Act was passed. The law had little impact on the lives of people in the lower socioeconomic statuses, but did grant suffrage to some small landowners, tenant farmers, and shopkeepers. Among the men and women of Lady Byron's acquaintance, preference for charity and political stability coexisted with the embrace of the new in science and technology. Victorians eagerly anticipated the improvements to life, but had limited solutions for the havoc they would bring.

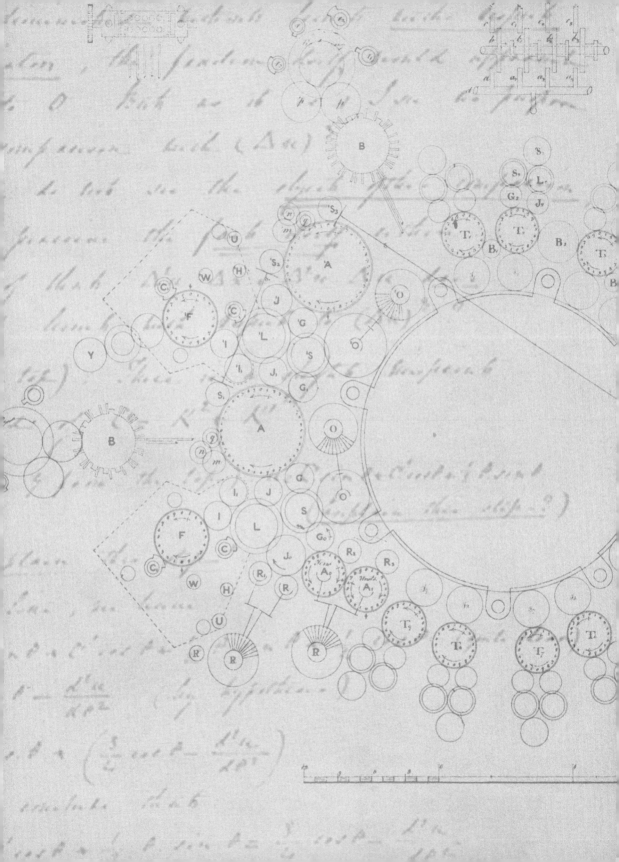

Part Three

1833–1835

Chapter 7

Meeting Babbage

Though Ada Byron was brought up very differently from other aristocratic girls her age, marriage was considered the main goal for her as it was for all young women. Lady Byron was a serious philanthropist and reformer, not a social butterfly, but her daughter had to find a husband nevertheless. That meant introducing her to society.

She came of age in a more reactionary social hierarchy than her mother and grandmother had known. A few decades before, everyone in the aristocracy knew everyone else, and a wide range of reckless behaviors was tolerated among the Old Guard, particularly if it was amusing. As the Industrial Revolution made huge new fortunes for factory owners and merchants, these nouveaux riches clamored to be included in fashionable society. The aristocracy's gatekeepers enforced strict rules of etiquette and morality in order to defend the traditional class order, which put them near the top. The rules were much more stringent for women than for men.

On May 10, 1833, at age seventeen, Ada Byron, along with a flock of other young women of marrying age, was presented to King William IV, Queen Adelaide, and their court.

Ada had been hidden away all her life in country houses and was very ill for a long time, with little practice conducting herself in polite society. Ada's mother had understandably been apprehensive about what her daughter might do under the strain (or the provocation) of royal formalities. Lady Byron proudly reported that her young lioness behaved tolerably well and was courteous to the various dignitaries in attendance.

Frivolities were definitely not what Ada was about. The obligations surrounding her debut did not interest her very much.

It was a private party that Ada and her mother attended a month later, on June 5, 1833, that pleased Ada more "than any assemblage in the *grand monde.*" One of the guests in particular sparked Ada's interest: Charles Babbage, a widower a year older than her mother. He was a famous inventor, philosopher (as scientists were then called), and mathematician and held Isaac Newton's chair at Cambridge University.

That evening, Babbage tried to charm Ada's mother, who remained frosty. But Ada's bright, eager face and discerning questions gratified and intrigued him. They had much in common. At her age, he too had been passionate about algebra and calculus. He enjoyed playing word games, making puns, and inventing and breaking codes. He was trying to create a universal language.

Babbage invented a speedometer, an ophthalmoscope, which

Charles Babbage at age forty-one. He held the most prestigious chair in mathematics at Cambridge, and his salon drew curious and influential Londoners.

allows people to see the interior of the eye, a time clock for factory workers to punch in, and a device to monitor earthquake shocks. He created a scheme to organize the first British postal system. Eccentric and original to some people, he was considered a crackpot by others. When he traveled, he took along sets of die-stamped gold buttons to give to anyone who might happen to save his life — a likely occurrence, since he often risked it. He believed that everything ever uttered by man or woman persisted in the atmosphere, that the air itself was one vast audio library.

For all of his eccentricity, Babbage had the very serious ambition to advance the age. Machines were replacing manual labor everywhere. Now Babbage was working on a machine that was intended to replace *mental* labor. He had poured his energy and money into a calculating machine he called the "Difference Engine," capable of producing limitless tables of numbers. Such tables were necessary for navigation, for predicting life expectancy for insurance policies, statistics, and more. Calculations made laboriously by hand inevitably contained mistakes that could sink ships and ruin businesses. Babbage's machine, on the other hand, would perform them automatically, infinitely, and without errors. Mechanical calculators had been built before, but only his was automatic, completing a series of steps on its own once it was cranked.

Ada was thrilled by the audacity of his ideas. His mind was as quirky as her own. Before the evening was over, Lady Byron had agreed to accompany Ada to a demonstration of Babbage's Difference Engine. Only a small section of it had been built, but that was enough to demonstrate its potential.

Babbage's regular Saturday salon, or party, at 1 Dorset Street might attract as many as two hundred people, each of those guests considered a person of accomplishment, rank, intellect, or beauty. Ada's first visit to Dorset Street took place on a quieter Monday. She and her mother entered a room displaying a collection of silver automatons. A few other guests had already gathered.

Babbage led his visitors to a special dust-free chamber, which contained a small section of his Difference Engine. The model, of polished wood, brass, and steel, was beautiful, and so was its operation. To produce accurate numerical tables, it needed to perform only a series of additions, as when instead of multiplying 7 x 6 to reach 42, one adds 7 six times. The method of differences can be illustrated by taking a series of whole square numbers: 1, 4, 9, 16, 25, 36, 49, 64, etc. Subtract each one from the next to create a new series, the First Differences: 3, 5, 7, 9, 11, 13, 15, etc. Subtract each one of those numbers from the preceding one to get Second Differences — all equal to 2. The machine advanced from one column to another by adding differences. The first result calculated the next one, and that one the one following, and so on. If the last column was accurate, it meant all the preceding ones were too. That made proofreading long tables unnecessary.

When completed, the machine would have had twenty-five thousand parts, including the printer. Parliament had given Babbage a grant in 1823 to build the machine, but its thousands of parts, many of them identical, were precision-made by hand, using tools that also had to be invented and fashioned for their specific purposes.

Babbage had run through the grant money and his inheritance

This partial model of Babbage's Difference Engine was displayed in a special room in his London home. Ada saw it there when she was seventeen, an encounter that set the course of her life.

trying to build all the tools and parts of his engine. When he failed to pay his contractor, the man ran off with the tools and many of the drawings and held them for ransom. Work on the Difference Engine was therefore at a standstill until Parliament sent more funds, which it declined to do.

On that evening, Babbage cranked his engine with a dramatic flourish. Tall stacks of geared wheels turned, each tooth bearing a number, making ripples, like waves breaking on a shore, retreating and breaking again and again, across successive columns, as information was being visibly transmitted. Mathematical computations were performed mechanically. The last step was to print the results.

Babbage coyly called his engine's startling results "miracles," by which he meant that they were statistically possible but highly improbable. One witness, quite baffled, asked if the machine could still make correct calculations if erroneous figures were entered. (The engine does not know that the initial numbers are incorrect; it will make calculations using the numbers it is given, which is not the same as the engine calculating a desired answer using incorrect numbers.) Babbage was used to that sort of question. What was more interesting to him that day was Ada's reaction. She clearly had a grasp of the engine's principles.

Sophia Frend, who was present at the demonstration, made this observation: "While other visitors gazed at the working of this beautiful instrument . . . Miss Byron, young as she was, understood its working and saw the beauty of the invention."

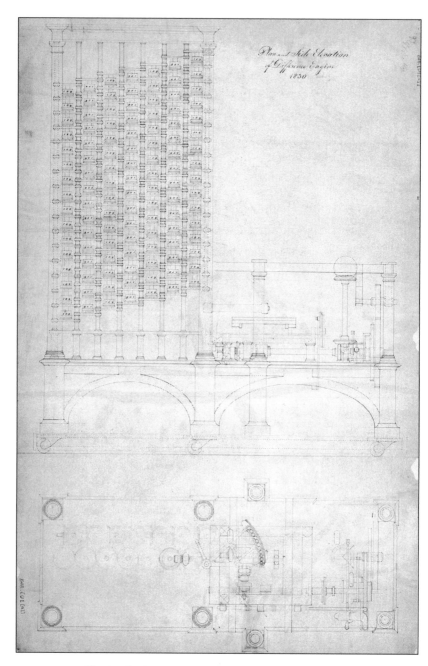

This detailed diagram of Babbage's Difference Engine dates to around 1830 and may be one of the drawings Ada studied.

Lady Byron had decided that Babbage's pursuit was merely "the whim of the moment." Her daughter was utterly captivated.

The Difference Engine was much talked about outside Babbage's salon. In the spring of 1834, Dionysius Lardner, a popularizer of new ideas in science and author of a book on Euclid (which Ada read), gave a set of public lectures on the Difference Engine at the Mechanics' Institute. Ada attended at least one of them and afterward asked Charles Babbage to show her drawings of the engine, which she studied carefully before pursuing her friendship with him. Did Ada realize that she now had an outlet for the coiled power of her imagination?

Chapter 8

A Role Model

It was probably Charles Babbage who introduced Ada to Mary Fairfax Somerville. An acclaimed mathematician and practicing astronomer, Somerville would take Ada, thirty-five years her junior, under her wing. When Mary was ten years old, her father, William Fairfax, had decided that she was uncontrollable and sent her to boarding school to be reformed. She became a voracious student and taught herself geometry. Later she recalled being "intensely ambitious to excel in something, for I felt in my own breast that women were capable of taking a higher place in creation than that assigned to them in my early days."

Mary Fairfax's first husband was Captain Samuil Samuilovich Greig, a Russian consul in London who disapproved of her studies. Nevertheless, she defied him and pursued them. After Greig died, Mary married Dr. William Somerville, inspector of the Army Medical Board and a cousin. Unlike Greig, he encouraged her studies, and Mary launched her mature work with a translation of *Mechanism*

Mary Somerville was one of the most eminent woman scientists of her day. She translated scientific works from French into English and made complex ideas accessible to the general public.

of the Heavens by the great French mathematician and astronomer Pierre-Simon Laplace.

Significantly, Somerville went beyond translating Laplace's text by adding her own thorough explanation of Laplace's mathematics, which was not familiar to his counterparts in England. As she put it, she translated his algebra into common language.

Somerville's next book, *On the Connexion of the Physical Sciences*, published in 1834, defined the study of the physical world in terms of the modern sciences that were then emerging — biology, chemistry, and geology. In her book's dedication to Queen Adelaide, Somerville said she wanted to make the laws of the universe available to her countrywomen. Written crisply and engagingly, it clearly explained what scientists knew then. In a review of Somerville's book, polymath William Whewell coined the word *scientist.* The book remained the top bestseller in England until Darwin's *On the Origin of Species* overtook it some twenty-five years later.

Mary Somerville was an exceptional scholar. She never claimed genius (wrongly believing it was not granted to females) and even coyly denied that she possessed originality. She was elected into the Royal Astronomical Society due to her great fame and her reputation as a modest, pleasant woman who didn't openly compete with men.

The Somerville household was Ada's first experience of a family. She often spent the night there after she and Mary had attended a Babbage salon or a scientific lecture together. One evening in November of 1834 at Mary's house in London, Ada, Lady Byron, Babbage, and Somerville's son, Woronzow Greig, were all lamenting the government's refusal to support Babbage's work. Mary Somerville wondered if the world were ready for such a machine as the Difference Engine. It was a good question. The government had lost faith in it, and few seemed to understand its potential.

Babbage replied that if he didn't build it, someone else would. A new government was being formed, and he hoped it would decide to fund his machine after all. Then, his eyes dancing, he told them in elliptical terms that he was designing a new, far more important engine — the *Analytical Engine.* Years earlier, Babbage had understood that the time and labor required for increasingly complex mathematical formulations would hold back scientific advancement. He felt that a way had to be found around the "overwhelming incumbrance of numerical detail." His answer was that the Analytical Engine, the inspiration for which had come to him like "throwing a bridge from the known to the unknown." He added, "It occurred to me that it might be possible to teach [a] mechanism to accomplish another mental process, namely — to foresee." This was an idea in keeping with the age of machines, but it seemed to take mechanical possibilities much further than anyone ever had. He told the group that he believed his new machine would guide the future course of science. (Babbage didn't actually call his machine the Analytical Engine until July 1841 — in a letter to German scientist Alexander von Humboldt, in which he complained that his invention was too advanced for England to support.)

In her journal, Lady Byron scoffed that his ideas were "unsound and paradoxical." But Ada was struck by their "universality." For her they had the potential he claimed for changing everything. The promise of invention, of the utterly new, thrilled her.

Ada and her mother toured the factories that had sprung up in northern England the summer before. It may have been at Babbage's suggestion. They wished to better understand the conditions of workers and view the latest technology. The pair inspected printers, ribbon makers, and textile plants. They especially admired the Jacquard looms, which wove complex patterns on fabric automatically. Lady Byron drew a picture of one of the punch cards that governed its operation.

Ada wrote to Dr. King that the machinery they had seen reminded her of Babbage's "gem of all mechanism," referring to his Difference Engine. Thousands of punch cards were used to program a Jacquard. An artist first created an intricate fabric design, then specialists punched "holes in a set of . . . cards in such a manner that when those cards are placed in a Jacquard loom, it will then weave upon its produce the exact pattern designed by the artist." The weaver could choose from different threads and colors, but the form of the pattern would always be the same. Each hole in a punch card corresponded to a hook that raised or lowered the harness, guiding the warp thread so that the weft thread lay above or below it. The cards stored information that could be rearranged and reused to make new patterns.

The Jacquard loom was an ingenious invention that allowed for automatic weaving of complex patterns using punch cards.

Chapter 9

Courtship and Marriage

Ada had already attracted at least one unsuitable fortune hunter who "made a very daring attempt to inveigle her . . . into a marriage because she was an heiress." Lady Byron, of course, was beside herself over the incident. Ada was still repenting for her "elopement," writing to Dr. King's wife in Brighton that she was far from satisfied by the "imperfection of [her] actions," adding that her feelings for her mother, while improved from a year ago, were "far from what they should be."

She pledged to avoid all "ordinary music" and the "exciting tendency of the pursuit & the display it might lead to," and to avoid all operas, plays, and balls (save the royal one).

Ada's luck changed when Mrs. Somerville's son, Woronzow Greig, introduced her to one of his classmates from the University of Oxford. William, Lord King (no relation to Dr. King), was a young diplomat who had recently come home from Greece to take charge of his family's estate. He was looking for a wife, and Woronzow, who admired Ada, told him that he knew an excellent candidate. In the spring of 1835, Ada and William found themselves together at a house party in Warwickshire.

William King fell hard for Ada, for her charming impulsivity, her brilliance, and her famous father. She was vivacious, outgoing, and spontaneous; he was dignified and shy, and their personalities seemed to be a complementary fit. Within days, they were engaged. Ada told no one but her mother. William may have been an unexciting aristocrat devoted to engineering projects and crop rotation, but he was by all accounts kind and decent.

Lady Byron was highly pleased with William — he had a title that was more than one hundred years old and owned several properties in Surrey, Somerset, and London. He was handsome and rich, and he sincerely admired Ada's intellect. The philosopher John Locke was his forebear, and his family estate had once belonged to William of Ockham, famous for "Occam's razor," the idea that the best answer is usually the simplest one. Most important, William admired and respected Lady Byron and would go to any lengths to please her.

Ada never worked up an ardor to match William's, but she was grateful for his affection and his decency. William wrote to her, "Since we parted I have been in [a] state of continued intoxication. . . . I look upon such happiness as too excessive to be enjoyed otherwise than in a dream." Ada replied more coolly that his letter was "an unexpected happiness" and that she felt "calm."

Marrying William allowed Ada to escape her mother's (and the

Furies') unceasing vigilance. An acquaintance teased Ada for so eas-ily surrendering the freedom she had looked forward to when she came of age. But marriage was inevitable, and William was a good man with modern ideas about women. He very much valued Ada's "enviable calmness and philosophy," her intelligence, and even her independence, and would give her leave to pursue her own interests, however unconventional.

The press got wind of the wedding. To avoid the curious crowds that would swarm a church, the ceremony took place at Lady Byron's home on July 8, 1835. Reporting on the event, the publication *World of Fashion* dramatically declared that Ada "has escaped the dangers and woes which enclosed her footsteps, the clouds that gathered round the morn of her life."

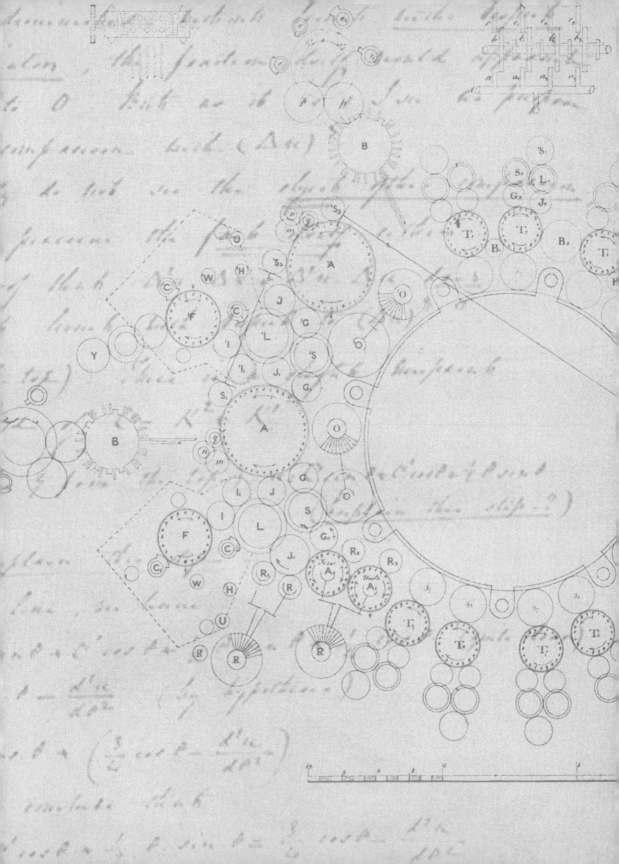

Part Four

1836–1842

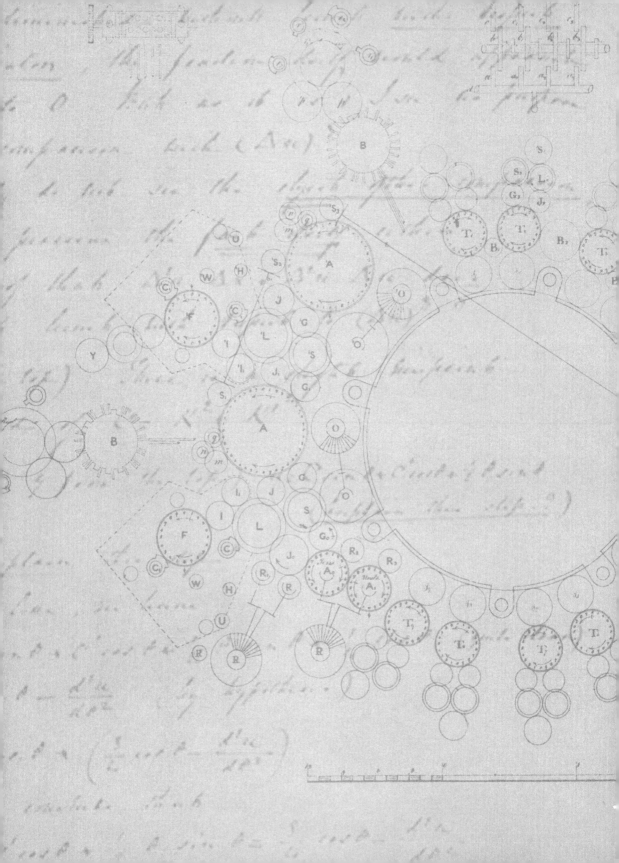

Chapter 10

Motherhood

Now that her domestic life was settled, Ada again threw herself into mathematics. She assured Somerville that her studies in trigonometry and in cubic and biquadratic equations continued. "So you see," she said, "that matrimony has by no means lessened my taste for those pursuits." She peppered Somerville with questions and received kindly replies. She asked where she might find wooden models to help her visualize and better understand the postulates of spherical geometry.

Lady Byron had slyly hoped that Ada would "prove the Mrs. Somerville of Housekeepers!" and "produce a Treatise 'On the Connexion of the Domestic Sciences.'" It was a joking reference to Ada's untidiness and Somerville's most famous work, *On the Connexion of the Physical Sciences.* Lady Byron delivered advice on every aspect of married life. That she herself had been married only a few months didn't inhibit her any more than it had inhibited the Furies, who never married at all. (The Furies wrote to Ada that she

Ada complained that the painter of this 1836 portrait had overemphasized her jaw, making it look large enough for the word mathematics *to be inscribed on it.*

should be grateful to William for marrying her, along with page after page on how to conduct herself as a wife.)

Ada's marriage had a remarkable effect on her mother: Lady Byron was suddenly *happy*. "I have made a discovery," she wrote to her daughter, "that in consequence of your marriage I am become more amiable, affable, and all sorts of pretty epithets, in the opinion of those who never thought me so before!— Such is the result of my Henship!" Lady Byron embraced William's admiration and respect, and they forged a powerful bond. His relationship with his own mother had been painful and Lady Byron filled an emotional void. He was her "Dearest Son." Together, they assumed responsibility for Ada's welfare.

Very soon after Ada was married, she was pregnant. During her confinement, Ada tried to keep up her mathematics, but there were far too many competing demands and distractions. For example, she did confess playfully to William that she had placed the odd bet at the racetrack, a pastime they shared.

Ada became a mother for the first time on May 12, 1836, at her home at 10 St. James's Square in London. Lady Byron suggested that if the baby were a boy, he be called Byron to perpetuate the name. Her association with the poet had been a personal torment, yet it was also a source of pride and she never let anyone forget about it. Ada followed her mother's suggestion — and enjoyed keeping a diary of little Byron's development.

She was pregnant again within months. To facilitate young Byron's weaning, he was sent to stay with the maternal Mary

Somerville. Ada wrote to Somerville to say how much she missed her son and imagined the "little treasure" must be crawling. She hoped the new baby would be another boy, for a girl would never "leave me in peace."

Mary Somerville rather incongruously praised Ada for managing her family and studies so well; she was "just the person to have a large family and I congratulate you on being in the fair way for it." (Somerville had six children but didn't undertake her scientific work until they were grown. Ada struggled with trying to balance work with motherhood and was temperamentally much more suited to the former.)

When Ada turned twenty-one, she was given a telescope and for Christmas the portrait of Byron that had hung behind the green curtain in her grandparents' house: a bold life-size painting of the poet at his most romantic, in Albanian peasant dress. The library at Ockham, where she lived with William, also contained volumes of Byron's poetry. Ada denied having much interest in poetry, probably because her mother was given to versifying, and so began thinking about how differently she might carry on the Byron legacy. She had herself painted in costume, too, as a Spanish dancer.

After her daughter, Anne, was born, Ada caught what she believed was cholera, a violent, often fatal, and untreatable disease caused by bacteria in sewage. Ada recovered and was soon pregnant with a third child. This wove Lady Byron more than ever into the fabric of her family as she offered herself as an expert on any problem. William didn't object. He admired but was daunted by his mother-in-law. In fact, he was so taken with her industrial and agricultural schools for children of the working poor, he was moved to set up one of his own. Ada participated by composing the curriculum and designing a gymnasium equipped with ropes like those she had used as a girl.

Soon there was yet another claim on her time: William was appointed Earl of Lovelace by the new young queen Victoria in 1838, making Ada a countess, henceforth to be addressed as Lady Lovelace, the name by which she is known today. She was obliged to attend a slew of balls, receptions, concerts, and the opera and was on the short list to become a lady-in-waiting. Fortunately, she was not picked. Had she been, her duties at court would have claimed most of her time and she could hardly have developed a reputation as a mathematician.

An avid follower of the latest advances in science, Ada went to hear a lecture on electromagnetism by Michael Faraday. Faraday was the rare scientist in Victorian England who was not an aristocrat and had educated himself in part by attending public lectures. His work was thrillingly of the moment. He had recently discovered electromagnetic induction, which led to his inventing the electric transformer and electric generator. Until then, electricity had been a curiosity. Faraday made it the next great technology, and it fascinated Ada.

Women could and did attend lectures, especially if the topic was popular in society. It was in such settings that Ada also saw demonstrations of the telegraph and the divining rod. But she was no passive observer. She was preparing to make her own contribution to science.

Doing independent scholarly work while married with small children and keeping one house in London and two in the country (Ockham Park in Surrey and Ashley Combe in Somerset) was very

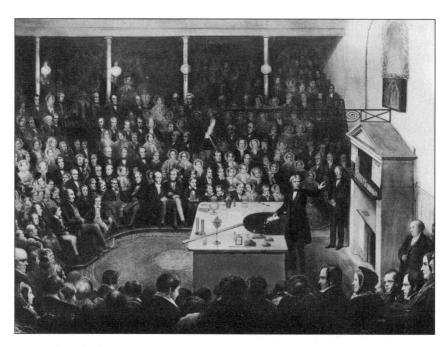

Michael Faraday, inventor of the electric generator, was a popular speaker of the time.
Ada was fascinated by his work and attended his lectures.

hard to accomplish. Even so, Ada was able to maintain her friendship with Babbage by mail when she was not in London. Knowing that he enjoyed games and probabilities as much as she did, she suggested in one letter that the language of math could be used to write a formula that would consistently win at games of chance. Ada was imagining a language of symbols for beating the odds; it was a radical notion and contained the seeds of her future achievement.

Chapter 11

Grasping for Mathematics

Ada lost her great model, confidante, and mentor in 1838 when Mary's husband became ill and the Somervilles moved to more affordable Italy. Ada asked Babbage to recommend a new tutor so that she could keep moving ahead in mathematics. "I have a peculiar *way* of *learning,*" she wrote, "& I think it must be a peculiar man to teach me successfully," hinting that Babbage himself could teach her. "You know I am by nature a bit of a philosopher, & a very great speculator. . . . Am I too imaginative for you? I think not." He didn't rise to the bait, though she had said tutoring would be "the greatest favour any one can do me." Babbage did make discreet inquiries about a tutor for her (without disclosing her name) but could find no one suitable.

Ada also wanted to be useful to Babbage; she and William shared a great faith in his Difference Engine, which still had not been built for lack of funding. (Skeptics joked that the machine should be asked to compute when, if ever, it would become useful.) Babbage continued to work on the design of his Analytical Engine. He told Ada he had thought of improvements that set his drawings back six months.

It was Lady Byron who found her daughter a new tutor. He was Augustus De Morgan, a member of the Analytical Society, which Babbage had cofounded as an undergraduate at Cambridge to modernize the British study of calculus. Augustus was a professor of mathematics at the University of London and had married Sophia Frend.

Augustus De Morgan, one of Ada's tutors

Ada began a correspondence course with Augustus, and they met in person once a month in London. Ada had specifically requested tutoring in "the Algebra; — Trigonometry; — . . . Differential Calculus; — & mere practice In Differentiation." The first book on modern algebra had been published in 1830 and presented algebra as more than a way of manipulating mathematical expressions, but a discipline of its own involving relations between symbols.

Differentiation referred to the method of finite differences, in which, Ada happily told Babbage, "I have more particular interest, because I know it bears directly on some of *your* business."

Augustus was not paid, except in the occasional gift, and Ada was always conscious of robbing time from his work. Modern mathematicians note that her letters to him are filled with questions that seem to reveal more confusion than comprehension. Yet that doesn't necessarily mean she was a slow student. Ada was a perfectionist and might not have written to him about ideas she easily grasped, but only those that challenged her. She complained of being flummoxed by logarithms, for example, yet was able to catch mistakes in the textbooks Augustus gave her. Some of her confusion must have been due to outdated or erroneous information given to her by Dr. King. To Lady Byron she reported, "The Professor & I certainly pull well together. Never was a better hit than that. — It will be a *slow* business, but a very certain one I imagine."

The episode of cholera seems to have worsened the nervous ailments that had plagued Ada most of her life. She complained of asthma, gastritis, hysteria, and what she called "heart attacks." Once she asked

Sophia to explain to her husband that ill health, not indifference, had made her fall behind in her studies. Although Ada begged her not to, Sophia relayed that message to Lady Byron, who in turn told Augustus that while he might be inclined to ease up on Ada's lessons because of these complaints, she advised against it. Ada's well-being, along with her happiness, depended on her progress in science. Lady Byron was an extremely active person, despite her hypochondria, and she expected Ada to follow her example in overcoming whatever conditions threatened to hobble her. Notwithstanding her habitual criticisms, she was ambitious for her remarkable daughter.

Augustus De Morgan believed that had Ada been a student at Cambridge University with such "power," she would surely be "an original mathematical investigator, perhaps of first-rate eminence." He even "prophesied" that Ada would go on to accomplish far more of real importance than even Mary Somerville had. Ada was a much more adventurous thinker than her older friend. But he believed Ada's body was rebelling against her intellect and told Lady Byron that he refrained from giving Ada too much praise for fear it would encourage her to overdo. Lady Byron noted tartly that if Ada "would but attend to her stomach, her brain would be capable even of more than she has ever imposed on it."

Ada wrote that she was on a mission, imagining that she perceived things no one else perceived, "things hidden from eyes, ears & the ordinary senses," her "immense reasoning faculties . . . the power . . . of throwing [her] whole energy & existence into whatever" she chose. "I can throw *rays* from every quarter of the universe into *one* vast focus."

This intellectual drive and emotional intensity caused her to soar and crash, soar and crash. She imagined mastering many branches of knowledge and music. Her Byronic impulses had been suppressed, while everything mathematical was encouraged. Now it appeared that she was trying to bring those two sides of her nature into productive harmony.

Lady Byron enlisted a reformer friend, Dr. James Kay, to treat Ada's chronic stomach pain and asthma. He prescribed laudanum, the mood-altering and addictive drug derived from opium. Physicians had been prescribing this drug freely for decades, especially to women. It eventually became the cure-all ingredient in over-the-counter medicines for complaints from coughs to depression. Ada became a regular user, and it seemed to calm her.

She didn't concentrate only on mathematics but continued to flit from one interest to another, playing her harp for hours at a time, singing, and, as the years of marriage rolled on, openly flirting with men at social gatherings, according to gossip. She was Byron's daughter, and people noticed. Her antics appeared in a newspaper society column: "The resemblance of Lady Lovelace to her renowned father, beyond some parental likeness, has as yet been confined to a certain amount of eccentricity. Her ladyship from her whimsical notions is thought a little daft."

Chapter 12

The New Engine

The British government still refused to fund his inventions, but Charles Babbage continued to make drawings, notes, and models for his Analytical Engine. He gave up all other pursuits, even while assuming it might never be built. "I am myself astonished at the powers I have given it," he wrote Ada.

She eagerly offered her assistance: "*My head* may be made by you subservient to some of *your* purposes & plans." But the opportunity to help him was not yet at hand.

In 1839, one of Babbage's friends presented him with a masterpiece of the Jacquard technology. A small portrait of the inventor Jacquard himself had been mechanically woven in silk thread and was so detailed and lifelike that it could hardly be distinguished from an engraving. Babbage loved showing it to people, asking what they thought it was and delighting in their astonishment when he told them. It had been created on a loom using twenty-four thousand punch

Babbage delighted in asking friends how they thought this lifelike portrait of Joseph Jacquard was created. None guessed that it had been woven on a Jacquard loom.

Punch cards like these would have enabled the
Analytical Engine to perform almost any calculation.

cards and was, in effect, a digitized image. On seeing this clever creation, Babbage realized that he could replace the revolving numbered drums he had envisioned for his Analytical Engine with punch cards.

The Analytical Engine was designed to be much more than a calculator. The punch cards, programmed like those in the Jacquard loom, would enable the machine to perform many operations (including addition, subtraction, multiplication, and division) and to store numbers and intermediate values. Unlike the Difference Engine, the Analytical Engine would perform almost any calculation (just like a modern computer) because the punch cards contained versatile programs that the machine could read, whereas the calculator used only mechanical parts whose operations were limited.

The dream of automation captivated many of Babbage's contemporaries. John Clark of Bridgwater had invented an apparatus he called the Eureka Machine, which automatically generated Latin verses. (He also invented a more useful item, a type of waterproof fabric.) Sir Walter Scott, the prolific novelist, declared that his Waverly novels had been written as if on a loom. The Romantic poet Samuel Taylor Coleridge also explored the idea of automatic writing, describing words as stereotypical symbols and comparing them to the notes on an automatic barrel organ. Ada's own father had referred to words woven into song. In addition, the human mind was seen by many people as so marked by experience that nothing truly original could emerge from it — a mind merely played out what it had absorbed in ever new configurations.

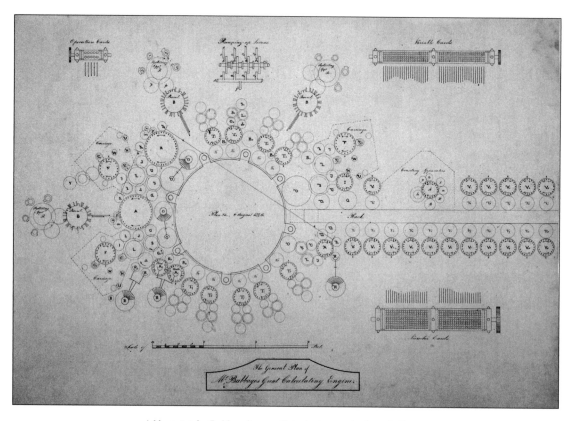

A blueprint for Babbage's great Calculating, or Analytical, Engine

It was in this environment that Babbage had conceived a machine — the Analytical Engine — that would ingeniously combine and recombine numbers, store them in its memory, and then decide what to do with them without any further human intervention, as if it were a giant computer chip. Lady Byron had called his Difference Engine a "thinking machine," a term commonly used then, but this new engine was more deserving of the name.

* * *

It wasn't until Ada was twenty-five, in early 1841, that Lady Byron finally revealed the specifics of Lord Byron's crimes to her. (Over the years, she had confided those unmentionables to many others, carefully constructing a version of events that emphasized her own virtues.) Byron, she claimed, had sired a child by his half sister, Augusta. The child, named Medora, was now living impoverished and neglected in France. Lady Byron had decided to undertake her rescue and redemption.

Ada replied that she was not astonished by the news, having long had her own suspicions. But she sensibly pointed out that since Augusta was married to someone else at the time of the supposed conception, it would be difficult to prove Byron's paternity. She hoped to learn sometime "how you ever came even to suspect anything so monstrous. The natural intimacy & familiarity of a Brother & Sister certainly could not suggest it, to any but a very depraved & vicious mind, which yours' decidedly was not."

The new disclosures encouraged Ada's lofty interpretation of her legacy. "If my poor Father had but possessed a little of my real philosophical turn!" she wrote to her mother, assuring her also that her own natural character was cheerfulness, when occupied and healthy. She declared that she would make amends to the world for her father's crimes: she had "an ambition to make a *compensation* to mankind for *his misused* genius." Possessing "less flash" and "more depth" than he, she would use the genius she had inherited from him to "bring out great *truths & principles.* I think he has bequeathed this task to me!"

Chapter 13

A Restless Student

A few months earlier, Babbage received a gratifying invitation from a friend, astronomer and mathematician Giovanni Antonio Amedeo Plana, to show his designs to a convention of philosophers in Turin, Italy. At last, interest in his invention! Babbage hurried to Turin with models, drawings, and notations. He received delegates at his lodging on a succession of mornings to describe his machine and answer questions.

Plana intended to take notes but turned that job over to a young engineer named Luigi Federico Menabrea. (This was a good thing, because Plana afterward concluded that the machine would be useless.) Babbage recalled later that having to explain the machine to delegates helped to order and refine his thoughts and insights. As the audience asked questions, he revised some of his design on the spot.

Menabrea wrote that Babbage had proposed "nothing less than the construction of a machine capable of executing not merely arithmetical calculations, but even all those of analysis, if their laws are known." Its operations would be given to it by "algebraical notation."

Babbage referred to the process as the machine "eating its own tail." The apparatus could use the results of its calculations to perform new operations, taking it far beyond fundamental adding, subtracting, multiplying, and dividing. The first set of punch cards fetched the numbers to be calculated, then the Variable Card selected numbers from among the Number Cards or from the memory, which Babbage called the Store. An Ingress Axis transferred the numbers to central wheels. An Operation Card then ordered an action — addition or multiplication or something else, this all taking place in the area called the Mill. Results were then looped to another group of cards, called Combinatorial, determined by the results of the first calculations. The results of the entire operation could be printed. If at any stage there was an error, the machine jammed, a bell rang, and a card shot up that read WRONG.

Even so, Babbage did not yet envision the machine's full marvelous potential. He did, however, realize that its size would be immense. If equipped with one thousand variables, for example, the machine would have to be five hundred feet long, fifteen feet high, and six feet wide.

On January 5, 1841, Ada wrote to Babbage, urging him to visit at Ockham, where she and William were living. She was eager to show him her "Mathematical Scrap-Book"— a volume that has not been found. And she wanted him to explain the "main points relating to your engine. — I have more reasons than one for desiring this."

The same day, she composed an essay exploring the force that was driving her. Imagination, she wrote, had two functions: "First: it is the *Combining* Faculty. . . . It seizes points in common, between subjects having no very apparent connexion, & hence seldom or never brought into juxtaposition. Secondly: it . . . brings into *mental* presence that which is . . . invisible. . . . Imagination is the Discovering Faculty."

Her essay continued: "Mathematical Science shows what *is*. It is the language of unseen relations between things. But to use & apply that language we must be able fully to appreciate, to feel, to seize, the unseen, the unconscious." While Babbage was finding new practical applications for numbers, Ada saw them as symbols with almost supernatural powers.

Mathematics made Ada cheerful and increased her self-confidence, as well as her determination to "add my mite to the accumulated & accumulating knowledge of the world."

She was also fascinated by the brand-new sciences of electricity, magnetism, and hypnosis, powerful forces operating beneath the surface of reality. Similarly, Ada was attracted to the brand-new field of photography and wondered if unseen phenomena could be captured by a camera.

Writing to her mother, Ada laid claim to qualities which equipped her to discover hidden realities of nature. She claimed perceptions that no one else had, along with the power to throw her whole self and energy into whatever subject or idea she chose.

Electrobiology and animal magnetism were in vogue in Ada's social circle. Mary Shelley's novel *Frankenstein; or, The Modern Prometheus* was partly inspired by the public attempts of Giovanni Aldini to reanimate corpses with electrical charges. Electricity was a thrilling and dangerous topic and widely discussed. It was feared that individuals and even groups could be controlled by others, were they to tap the electricity that resided in everyone. Ada gave it a great deal of thought, as well, deciding that the mode of action of electrobiology was the "bond of union between the *mind & muscular action.*"

Mesmerism, which appeared to activate an innate magnetism, was named for the Frenchman Franz Anton Mesmer, who claimed to cure ailments by magnetic hypnotism. Mesmer had first tried to draw magnetic powers from the stars but soon decided that the force he detected resided in his own person and that with it he could put his subjects into trances. A committee in the late eighteenth century that included Ben Franklin and Pierre-Simon Laplace assessed the scientific validity of the practice and declared it bogus. But it was later revived in England, where Lady Byron's friend Harriet Martineau claimed she was cured of her cancer by mesmerism. (And of her dependence on laudanum.) Ada wanted to test it scientifically.

A demonstration of mesmerism was held in 1841 at the Lovelace home on St. James's Square, London, attended by Charles Wheatstone and other scientific gentlemen. Ada, standing near the subject who had been put into a trance by the mesmerist, suddenly was overcome by "unnatural feelings mental & bodily" that were profoundly troubling. Afterward, she decided that she was involuntarily mesmerized that night and that "all my ill health had its foundations in that." At least she hoped so. If she was right, time could heal her.

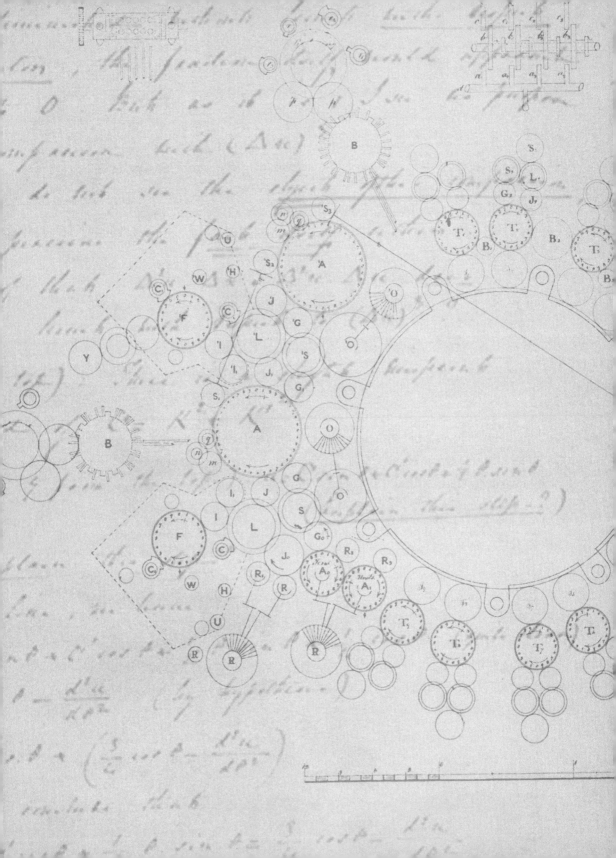

Part Five

1842–1852

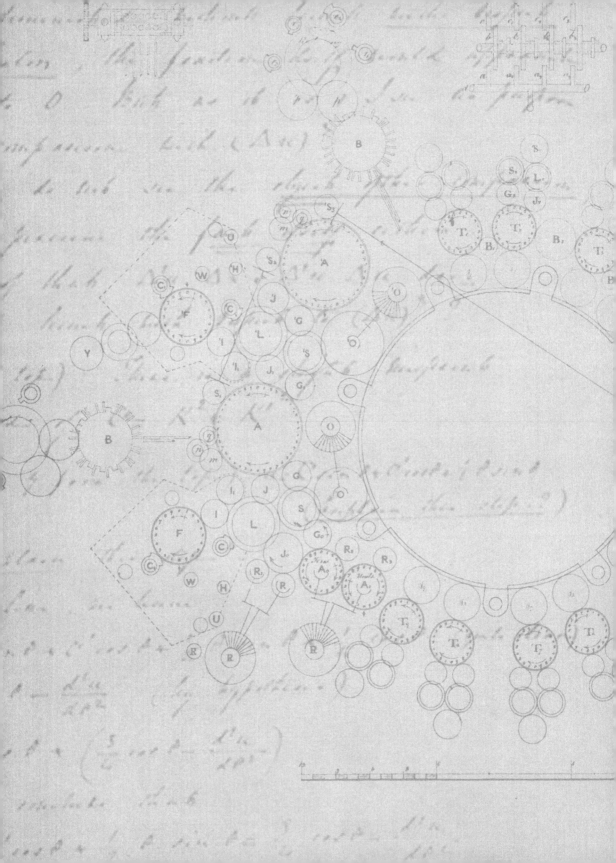

Chapter 14

Masterwork

In October 1842, a Geneva journal published the notes that Luigi Menabrea had taken at Babbage's Turin presentation of the Analytical Engine. Menabrea called the engine Babbage's "gigantic idea." His article caught the eye of Charles Wheatstone, who scouted foreign journals for material that could be translated for the publication *Scientific Memoirs.*

Victorian scientific and social circles overlapped. Wheatstone was well acquainted with Ada and her interests. He asked her if she'd translate Luigi's essay from the French for his journal. Ada was thrilled. A scientific translation had been how Mary Somerville first made her mark on the scientific world.

Ada went to work right away and apparently completed the translation even before she told Babbage she was doing it. She felt liberated and renewed. Babbage was very pleased by the result.

He asked Ada why she herself hadn't written something about the Analytical Engine, since she knew so much about it. She replied that it hadn't occurred to her. Ada didn't think the male-dominated scientific establishment would take an interest in her thoughts. But Babbage did. He urged her to add her own notes to the article. He still hoped, after his success in Turin, that the prime minister or Parliament would reinstate his funding. Only the government could underwrite a machine the size of a large locomotive. An essay by Ada — his muse, his interpreter, his colleague — might help persuade it to do so.

She knew her relationship with Babbage would benefit them both. Here was an opportunity to make herself as famous as Lord Byron, but for scientific achievement. Ada began enthusiastically enlarging on her translation. Not only did she comprehend what Babbage had invented; she was also uniquely qualified to explain it to others because of her learning in mathematics and science — but above all because of her singular imagination. Work on the essay soon consumed her.

She and Babbage sent letters back and forth between their homes by mail (delivered six times a day) and messenger (they lived a mile apart). They also met frequently. The balance in their relationship shifted as Ada's confidence grew. She teasingly scolded him for being too vague, slow, or careless.

Be kind enough to write this out properly for me.

You were a little harum-scarum & inaccurate.

I wish you were as accurate, & as much to be relied on, as I am myself.

My Dear Babbage. Unless you really have necessary *business to transact with me tomorrow, pray do* not *come to see me; for I am full of pressing & unavoidable engagements.*

Ada continued to suffer from stomach ailments and to take laudanum. Sometimes she had trouble breathing. Lady Byron's doctor bled her and advised drinking brandy to counter the drowsiness caused by laudanum. Still, her brain was working at a very high level, if sometimes in drug-induced spurts. Ada often worked through the night at Ockham, a King estate, or in London during her nine-month effort to master the complexity, and more important, the potential of Babbage's invention. ("I am doggedly attacking & sifting to the very bottom," she wrote. "I am grappling with this subject, & connecting it with others.") In the end, her notes grew to be three times as long as the essay she was commenting on, and full of original thoughts that anticipated the basics of computer programming.

Babbage had listed the operations the engine could perform, including ascertaining if a variable had a plus or minus value or zero, and multiplying without a table. Ada thought of numbers as more than quantities: they were symbols, and as such, they could stand for anything. Using them, Babbage's machine could be programmed to perform other operations beside mathematical ones.

And because "the fundamental relations of pitched sounds in the science of harmony" were subject to abstract operations, "the engine might compose elaborate and scientific pieces of music of any degree of complexity or extent."

"Many persons . . ." she wrote, "imagine that because the business of the engine is to give its results in *numerical notation,* the *nature of its processes* must consequently be *arithmetical* and *numerical,* rather than *algebraical* and *analytical.* This is an error. The engine can arrange and combine its numerical quantities exactly as if they were *letters* or any other *general* symbols."

This was Ada's great leap of imagination and the reason we remember her with such admiration. Her idea that the engine could do more than compute, that numbers were symbols and could represent other concepts, is what makes Babbage's engine a proto-computer.

"All this was impossible for you to know by intuition and the more I read your notes the more surprised I am at them and regret not having earlier explored so rich a vein of the noblest metal," Babbage told her.

Ada predicted operations known in modern-day computing, such as looping (a sequence of programming instructions that is continually repeated until a certain condition is reached) and selection (a sequence of programming instructions that takes one of two courses of action depending on the answer to a question, after which the program moves on to the next instruction).

In his autobiography, written many years afterward, Babbage wrote:

> *We discussed together the various illustrations [of operations] that might be introduced: I suggested several, but the selection was entirely her own. So also was the algebraic working out of the different problems, except, indeed, that relating to the*

numbers of Bernoulli, which I had offered to do to save Lady Lovelace the trouble. This she sent back to me for an amendment, having detected a grave mistake which I had made in the process.

Ada had written to Augustus De Morgan in 1841 about Bernoulli numbers — a sequence of rational numbers that appears often in mathematics and has many interesting arithmetical properties. Calculating them requires many sets of operations and was beyond the capabilities of the Difference Engine. Ada thought they were a good example of how an implicit function (a function in which the dependent variable is not isolated on one side of an equation; for

In a letter to Augustus De Morgan dated November 21, 1841,
Ada refers to her work with Bernoulli numbers.

example, the function $x^2 + y^2 = 1$ is implicit because the dependent variable y is on the same side of the equation as the independent variable x) could be carried out by the Analytical Engine. She asked Babbage for the necessary data and formulae.

Ada slogged through the data. It was a struggle, as she frequently reminded Babbage: "Before ten years are over, the Devil's in it if I have not sucked out some of the life-blood from the mysteries of this universe, in a way that no purely mortal lips or brain could do. No one knows what almost *awful* energy & power lie yet undeveloped in that *wiry* little system of mine." She was working like the devil for him, which perhaps she was (when not his fairy).

One day she complained of having gotten into such a "quagmire & botheration with these *Numbers*" that she gave it up to go horseback riding. She signed herself "Yours puzzle-pate."

But she didn't give up.

"I have worked incessantly, & most successfully, all day. You will admire the *Table & Diagram* extremely"— a reference to the program for computing Bernoulli numbers. "They have been made out with extreme care, & all the *indices* most minutely & scrupulously attended to." Ada eventually caught the mistake in Babbage's calculations.

The corrected result is in her famous Note G (the last of the appendix sections beginning with Note A). It lays out the steps for calculating the Bernoulli numbers, now referred to as the first computer program, taking the form of an algorithm, or series of step-by-step instructions for solving a problem.

To Note G she added a caveat for the reader: "It is desirable to

Diagram for the computation by the Engine of the Numbers of Bernoulli. See Note G. (page 722 et seq.)

This diagram shows the algorithm Ada devised for calculating Bernoulli numbers, which involves what we now call looping and batching. This algorithm is considered the first model of computer code.

guard against the possibility of exaggerated ideas that might arise as to the powers of the Analytical Engine. . . . [It] has no pretensions whatever to *originate* any thing. It can do whatever we *know how to order it* to perform. It can *follow* analysis; but it has no power of *anticipating* any analytical relations or truths. Its province is to assist us in making *available* what we are already acquainted with." (This observation is what inspired Alan Turing to propose "Lady Lovelace's Objection"—the idea that artificial intelligence amounted only to what a machine was told to do.)

She described a "science of operations," or what we call computing.

This science constitutes the language through which alone we can adequately express the great facts of the natural world, and those unceasing changes of mutual relationship which, visibly or invisibly, consciously or unconsciously to our immediate physical perceptions, are interminably going on in the agencies of the creation we live amidst. . . .

A new, a vast, and a powerful language is developed for the future use of analysis, in which to wield its truths so that these may become of more speedy and accurate practical application for the purposes of mankind than the means hitherto in our possession have rendered possible. Thus not only the mental and the material, but the theoretical and the practical in the mathematical world, are brought into more intimate and effective connexion with each other.

Ada had worked out her sketch solely in her imagination. There was no physical machine for her to experiment on. She wrote her visionary essay in pencil, so she could make corrections. Then, when she was finished, her adoring, admiring husband William inked it for her.

Babbage had met with the prime minister, Robert Peel, in November of 1842 to discuss whether the government would have further involvement in his Difference Engine. His timing could not have been worse. Peel was trying to contain rioting mobs of the starving poor whose jobs had been destroyed by technology. (The Luddites, championed by Lord Byron early in the century, had been the first to violently rebel against the mechanization of work.)

Babbage opened the interview with Peel by complaining that jealous rivals had poisoned the government's opinion of his machines. He told Peel that the Analytical Engine would transform all science — even all civilization. But he followed his claim with a litany of resentments. He was betrayed and hurt, and he wouldn't shut up. Finally, the exasperated Peel told Babbage that by his own admission, he had rendered the Difference Engine useless by inventing a better machine.

Babbage had failed to convince Peel of the importance of either of

This model of the Analytical Engine was built in 1871 and displayed in London's Science Museum.

his machines. Peel could not understand the point of building inventions that quickly became obsolete because their inventor had new ideas. To Peel, Babbage's engines were nothing but expensive toys anyway. "What shall we do," wrote Peel, "to get rid of Mr. Babbage and his calculating machine?"

Babbage was furious, of course. He commanded Ada to append an account of the government's ill treatment of him to her essay. She was appalled; these changes would make her essay a completely different document, and the changes would overshadow her work, which was supposed to be scientific, not an instrument for revenge. When he insisted, she called his fixation "suicidal." Suddenly, the collaborators were at odds. Babbage went behind her back to Charles Wheatstone, the publisher of *Scientific Memoirs,* while Ada was in the country. She exploded to her mother: "I am sorry to have to come to the conclusion that he is one of the most *impracticable, selfish, & intemperate* persons one can have to do with." She predicted that their relationship would cool ("He will never forgive me") but would not end in outright alienation.

There were other disagreements. Babbage thought Ada's notes were too important to simply be included with the translation, as if they were an afterthought. He wanted to ask Wheatstone to publish them as a separate work. Ada, who was eager to get her essay into print, protested that would violate her agreement with Wheatstone. It would certainly also delay publication. She prevailed again. The essay appeared with Menabrea's notes and without Babbage's grumbling about the government's treatment of him.

When the first proofs arrived with errors in July of 1843, it was

Ada's turn to be furious: "Out of *several corrections* made, not one is inserted; neither are the Upper Indices added; nor the little Foot-Note. . . . I cannot account for such negligence."

She also chided Babbage, "I do not think you possess half *my* forethought, & power of foreseeing all *possible* contingencies (*probable* & *improbable,* just alike)."

And she boasted, "I do *not* believe that my father was (or ever could have been) such a *Poet* as *I shall* be an *Analyst;* (& Metaphysician)."

"The *pithy* & *vigorous* nature of [my] style seem to me to be most striking; and there is at times a *half-satirical* & *humorous dryness,* which would I suspect make me a most formidable *reviewer.* I am quite thunder-struck at the *power* of the writing."

William had advised Ada to sign her essay with her initials, "A. A. L." (She was Augusta Ada, but never used Augusta.) She told Babbage, "It is not my wish to *proclaim* who has written it; at the same time that I rather wish to append anything that may tend hereafter to *individualise,* & *identify* it, with other productions of the said A. A. L." In other words, she expected to produce much more work for publication. And yet, she took William's advice to disguise her gender so that readers would take the essay seriously. The disguise didn't work. Too many people knew her, knew her interests, and knew her initials.

Two hundred and fifty copies of the essay were printed. Ada and Babbage distributed them to various colleagues and acquaintances of influence. De Morgan wrote to Ada that he found it very good. So did Mary Somerville. Joanna Baillie, a Scottish poet whom Lady Byron considered a friend, wrote belittlingly to Somerville, "The lady who we know so well as 'little Ada,' whose chief conversation used to be about a Persian cat, Puff by name, is beginning to be known a little in the literary world." Luigi Menabrea was given a copy by Babbage's son and praised Ada as a "savante" named "Lady Lovely."

Ada proudly sent the article to her mother, with a note: "No one can estimate the trouble & *interminable* labour of having to revise the printing of *mathematical* formulae. This is a pleasant prospect for the future, as I suppose many hundreds & thousands of such formula will come forth from my pen, in one way or another." A few days later, she added, "William especially conceives that it places me in a much *juster* & *truer* position & light, than anything else can. And he tells me that it has already placed *him* in a far more agreeable position in this country." With publication of her essay, recognition of her genius and a new career as a "formidable reviewer" seemed to be in hand. If the engine could actually be built, she and Babbage would change the world!

Ada persuaded William to stake their wealth on building Babbage's engine. Ada well knew how proud and independent Babbage was, but she believed that she could, with William's help, "bring his engine to *consummation,* (which all I have seen of him & his habits the last 3 months, makes me scarcely anticipate it ever *will* be, unless someone really exercises a strong coercive influence over him). . . . I shall be willing to be his Whipper-in during the next 3 years. . . . Much of this is W's suggestion." Ada and William sent Babbage a sixteen-page proposal for its production. Ada may have undermined her argument by accusing Babbage of loving fame, glory, and honors more than truth (while admitting: "Far be it from *me,* to disclaim the influence of *ambition* & fame. No living soul ever was more imbued with it than myself").

She and William would undertake to manage all practical matters, including raising funds and overseeing the machine's construction. As Babbage's interpreter, Ada would make certain the public knew and understood it. The cost of the enormous machine might well exhaust their fortune, but William loved and admired his wife. "What a General you would make!" he liked to exclaim.

Babbage angrily rejected their proposal. On her letter he scrawled "refused all the conditions." Like many a major investor, Ada was proposing to take possession of Babbage's idea and bring it to fruition, but on *her* terms. He could not agree — he insisted on being sole proprietor.

Chapter 15

At Loose Ends

Though they never collaborated on the building of the Analytical Engine, Ada and Babbage's relationship did survive the storm. Before long, Babbage was planning a visit to the Lovelaces at their home, Ashley Combe, taking with him "papers enough to enable me to forget this world and all its' troubles and if possible its' multitudinous Charlatans — everything in short but the Enchantress of Numbers . . . my dear and much admired Interpretess."

William was busy building towers out of the pink brick he made himself. The three Lovelace children continued to provoke Lady Byron by running steeplechase races on their beds, outwitting their nannies, and using, according to her, vulgar language.

Lady Byron nagged Ada relentlessly to find a tutor who could discipline them, while William favored sending young Byron, at least, to boarding school. Lady Byron held the opinion that her ex-husband had been corrupted by boarding school, and vetoed the idea. Ada, for her part, worried that she was neglecting the children, but unless someone were found to oversee them, she would have no time for metaphysics.

She had worked so hard on her essay, and what had come of it? Two hundred people had seen it, most without comprehending; that made it a mere vanity publication. Furthermore, the Analytical Engine itself seemed doomed. She had called the essay her firstborn and expected it to have many siblings in the course of what she had called her profession. Restless and unhappy, she was drinking claret, taking more laudanum, and eating very little. William proved oblivious to his wife's vulnerabilities as evidenced by a note to Lady Byron in which he said of Ada that people "must be struck with the grandeur & nobleness of her intellect — she has but to be natural to be as much loved as she is to be admired & wondered at."

Lady Byron again took charge and hired a rising young physiologist named Dr. William Carpenter to tutor the children and, as Lady Byron made clear, to manage Ada and her moods. Carpenter, appointing himself a kind of therapist, found Ada eager to confide not only her symptoms — upset stomach, depression, anxiety — but her innermost feelings too. She told him of her teenage elopement and admitted that her affection for William and even for her children had cooled. As Ada opened her heart, Carpenter found himself helplessly drawn to her and confessed, in turn, that his own wife's company was almost unbearable to him at times.

Acknowledging a dangerous mutual attraction, they met alone on a few occasions. By the time Ada suggested that they "check" their relationship, it was too late. William had learned of the charged meetings and tried to fire Carpenter. Lady Byron, however, insisted Carpenter remain in service for a few more months. Eventually even she abandoned Carpenter's cause and he was given a sum of money to go away.

As Ada and William continued to drift apart, Ada spent longer periods in London. She asked William to apply to the Royal Society for permission to borrow books and scientific papers from its library. William had been made a member, she believed, because of his Byron connection. Women were forbidden even to enter the building. (A bust of eminent mathematician and astronomer Mary Somerville had a place of honor in the lobby, but she, too, was barred from going inside.) Ada also begged Woronzow Greig to find out if she could slip into the library early in the morning, when no one else was around. She assured him the secretary was discreet and would not make a fuss. Despite her connections to the scientific establishment, her gender was a major handicap.

Serious work was also undermined by her chronically poor health. Insomnia, loss of appetite, exhaustion, then hunger and terrible headaches all lasted for months. To her great regret, she often had to shun society. Once, she wrote to Lady Byron that her doctor had cried, "Oh! At least you have lost that *mad* look." This physician had been prescribing various proportions of laudanum, porter, claret, and morphine, as well as bleeding. They brought only intermittent relief.

Chapter 16

Restless Spirit

Early in 1844, Ada developed alarming new symptoms: her face suddenly swelled; she was sleepless, too thin, and always cold, and often lost her balance. Stimulants and narcotics failed to relieve any of it. Her lower back ached and often she shivered helplessly. "Galvanization [electro-therapy] in *winter,* & sunbathing in *summer,* are the things for me," she told Lady Byron, in the perky tone she usually used with her mother. Her illness made her want to know more about the human body. Hers was an *"experimental laboratory . . .* inseparable from me." It made the body a natural subject for her restless intellect.

German research on the nervous system and the molecular composition of blood was a place to start. "A Newton for the *Molecular Universe* is a crying want," she wrote in a note. Could the brain and its functions be expressed in equations, laying out laws for mental activity, like those of gravity and the planets? "I hope to bequeath to the generations a *Calculus of the Nervous System,"* she told Woronzow Greig.

She asked Michael Faraday, the great pioneer of electromagnetism, if she could be his apprentice and replicate his experiments as a way to master the field. Then she would test electricity on the human nervous system. Faraday refused on grounds of his ill health, but otherwise his response thrilled her: "Faraday expresses himself in absolute amazement at what he . . . designates the *'elasticity* of my *intellect.'*" She boasted that he seemed to think her the *"rising star* of Science."

Charles Wheatstone, the inventor of the electric telegraph and publisher of Ada's translation and her notes, suggested that she try to persuade young Prince Albert to support the Analytical Engine. Wheatstone knew that Albert hoped to establish a scientific circle in England and had met resistance from the aristocracy, which generally disliked him. Perhaps Ada could even become Prince Albert's science advisor! But nothing would come of that.

Understanding that her ambitions could never be realized unless she found the means to conduct actual experiments, she sought out a reclusive Ashley Combe neighbor named Andrew Crosse, an eccentric who had turned his country mansion into a hazardous electrical laboratory, draping more than a mile of copper wire over the trees. He was working on the voltaic battery to capture electric charges and on the production of sound by electromagnetism. Neighbors called him the "thunder and lightning man." Crosse has been compared to the twentieth-century genius Nikola Tesla. Ada wrote to Crosse, referring to herself as "the bride of science," adding that science was religion to her and that religion was science. She claimed she could write a volume on the interconnectedness of everything in the universe and proposed a visit to observe his experiments so she could conduct her own explorations of electricity and the nervous system, using frogs.

The Crosse household impressed her as "the most *unorganized* domestic system I ever saw." Crosse had "the most *utter* lack of *system* even in his Science. . . . I have quite a difficulty to get him to show me what I want. *Nothing* is ever *ready.* All chaos & chance." As a result, Ada lost interest in electric experiments. She did, however, form an intense friendship with Crosse's son John. They talked long into the night about mutual interests. John was a lively foil, opposing every notion Ada put forth with excellent arguments, according to Ada. He became her frequent visitor in London and at Ashley Combe, which led to rumors of an affair.

When Woronzow warned Ada about the gossip, she dismissed it, saying everyone in society was slandered at times, adding, "My character can't be mended now. It was utterly *gone* before I was 26." She referred to her *"life-less life"* as *"one continuous & unbroken* series of *small* disappointments; — & has long been so." Of William she said, "It is not his fault that to *me* he is *nothing* whatever, but one who has given me a certain *social position."* Furthermore, her children were "irksome *duties."*

At times, she was depressed and irrationally anxious about everything, even thunderstorms. She wanted serious work, yet serious work required a stable temperament. Her father had bequeathed her a volatile one. Or was it the drugs she took that so agitated her?

Ada did begin an essay on the molecular structure of matter, but her growing passion for betting on horse racing had become a distraction. She shared with Babbage a fascination for theories of probability (he had written a book about games of chance). Some biographers think they may have collaborated on a system for betting on horse races, but there is no evidence of it. Ada was awarded only a meager allowance when she married, so maintaining a gambling habit meant borrowing money — more and more of it. At first, she and William had placed bets together and Ada hadn't felt she had to conceal them from her mother. Now that the amounts had increased in size and frequency, Ada's bets were discreetly placed by her maid, Mary, who had previously worked for Babbage.

It didn't take long for Ada to be drawn into a shadowy betting syndicate, one that included John Crosse. It was a dangerous association with untrustworthy partners, and she was in debt for a great deal of money. If her mother found out, Ada wrote to Woronzow, "it would do irreparable mischief in more ways than one." She and William had already borrowed from Lady Byron to finance his construction projects.

One friend thought Ada was overconfident that her knowledge of mathematics would give her an advantage over the bookmakers. (Ada's granddaughter later remarked to one of Ada's biographers that horses never would cooperate with mathematicians.) By 1847, Ada's debts were so great that she appealed to Woronzow for help and he arranged for a bank loan (something a woman could not do on her own). When Ada heard that Lady Byron was about to consult Woronzow on a legal matter, she frantically begged him not to breathe a word of her loans. She did finally have to reveal her debts to William when a bookmaker refused to be put off. Devoted as ever, William paid him.

As she neared the age of thirty, Ada wrote to Lady Byron that despite her sufferings, she was yet vigorous enough to do anything. Her ambition required her to learn more, but it wasn't too late to begin an education. She asked her demanding mother if they were now in agreement about poetry, music, and philosophy. "You will not concede me *philosophical poetry.* Invert the order! Will you give me *poetical philosophy, poetical science?*" It was her great gift to see that science was a field for the imagination.

But people who saw Ada in her early thirties were shocked by her visible decline and halting speech. At a London dinner party, she collapsed and then apologized to the host for "spasms of the heart." A close companion of Lord Byron, John Cam Hobhouse, sat next to her on another occasion. He described Ada as "poor thing" and wrote "she is looking very ill indeed . . . [but] she spoke to me very freely on subjects few men & no women venture to touch upon."

Even in poor health, Ada's playful instincts were easily awakened. She went often to Brighton for the healthful sea air, and noticed that the clumsy swimsuits women wore presented a hydro-engineering problem — their voluminous skirts ballooned when a swimmer entered the water and would suddenly upend her, head underwater and feet in the air. Ada designed a one-piece costume to solve the problem.

It finally occurred to her that there might be a simple cure for her distress. With a great deal of willpower, she stopped mixing wine and brandy with laudanum. The absence of this intoxicating and clouding mixture put her on a more even keel for a while.

The Lovelaces continued to entertain Babbage from time to time;

he came to their country house by train and was given a pony to ride around the grounds. His latest invention was an automaton that could play tic-tac-toe. He hoped to tour it to raise funds to build his Analytical Engine. Ada discouraged him, alarmed that it would fail: "I want you to *complete* something."

William and Ada tried to put him in touch with wealthy aristocrats who might make the building of his Analytical Engine a reality, but Babbage couldn't seem to help undermining his cause and turning off would-be investors. His offense was garrulousness, and a penchant for telling endless ribald stories in mixed company. In a letter to Lady Byron, William observed, "I hope Babbage may succeed — but one is in a minority when one advocates the philosopher who seems condemned to an orbit of his own — and this too by men who have met him at our house & have been struck with his genius. . . . [They] are rather plagued by his excessive fulsomeness. I do wish he would leave those things, which do not suit his age & exterior."

Ada and William were on an extended tour of northern England in 1850 when they visited Lord Byron's childhood home, Newstead Abbey, for the first time.

The owner had done extensive restorations on a property that even in Byron's time was in disrepair. The approach was at a lake where huge model boats had once waged mock naval battles. One half of the great stone edifice contained the repaired baronial hall, while the other consisted of the Gothic abbey's remnants. Byron's verses describing the place must have run through Ada's mind. She felt the weight of family history: "All is like *death* round one; & I seem to be in the *Mausoleum* of my race [family]. What is the good of living, when thus all passes away & leaves only cold stone behind it?" It made her afraid that she would die without achieving enough, since fate stood in her way. "We ought to have been happy, rich, & great. But one thing after another has sent us to the 4 winds of Heaven."

Newstead Abbey in Nottinghamshire, England, was Lord Byron's childhood home.

Their host, who had gone to school with Lord Byron, had prepared for the visit by reading up on the latest in science, and he eagerly welcomed the Lovelaces. To his dismay, he found Ada unresponsive and dull, "a perfect blank." Nevertheless, he trailed her doggedly the next day as she circled the grounds, ignoring him. They walked. Suddenly, she turned and began apologizing for her incivility, praising the improvements he had made and telling him what great meaning the place had for her. Later, he would declare he had "never before met with so agreeable and cultivated a lady." For her part, Ada reported that being there with the ghosts and their shrine made her feel that she had "had a *resurrection*." She now loved her Byron forefathers, despite her mother's efforts to alienate her from them, and embraced her role as their heir.

Ada wrote to Lady Byron about the delight she took in finally seeing her ancestral home. Her mother, who had not been told that Ada and William had planned a visit to Newstead Abbey, replied crossly. She cautioned against subscribing to a "Mythic idea" of Byron. Her great fear seemed to be that Ada's children would grow up thinking that she, Lady Byron, had coldly deserted Lord Byron. If they thought that, it would be "better for them not to have known me." Rather than condemn him, she was now claiming to have been Byron's "best friend."

William hastened to assure his mother-in-law that they were still on her side. But the mutually confiding, sometimes teasing relationship was fatally wounded. Lady Byron had endowed much of her life with two main themes: devotion to her daughter and martyrdom to Byron's wickedness. Ada seemed to want no more to do with them.

After the tour of northern England, the Lovelaces went to the races with friends. On that occasion, the horse Ada had bet on, the famous Voltigeur, won handily. Ada told Sophia Frend that she had never felt better. She now possessed a letter signed by William that granted her permission to gamble.

The following spring, all of England was caught up in the excitement surrounding the Great Exhibition in the Crystal Palace. It had

The Great Exhibition of 1851 was held in an enormous glass structure dubbed the Crystal Palace.

been conceived by Prince Albert to showcase the achievements of the Industrial Revolution in Britain and around the world. Babbage was in demand as an expert guide to important visitors, though his own Difference Engine was not on display. He did, however, place a demonstration model of an electric signal in a window of his house while the exhibition was in progress, and its flashing light won him some attention.

The Lovelaces visited the exhibition many times, marveling at the astonishing variety of inventions and products. It was a proud moment for Britain. William was even awarded a prize for brick making. But the feeling of celebration was cut terribly short.

In June, Ada suffered massive hemorrhages of the reproductive tract. All of her various complaints paled before this new manifestation of the undiagnosed condition that had been developing for several years. Her doctor could only offer vague reassurances, which Ada conveyed to William: "the *local* condition is no longer *vicious.* Dr Locock explained to me yesterday how threatening & how morbid it *had* been. . . . He said that tho' now there is still an extensive deep seated *sore,* yet it is a *healthy* sore." In an undated letter thought to have been written in late 1844 or early 1845, Ada had referred to symptoms of this disease.

Ada admitted to her mother that her illness took a toll on her mind. She couldn't concentrate, was often confused or "dulled." The visit to Newstead Abbey had a deeper significance for her. She wrote of "that horrible *struggle,* which I fear is in the Byron blood. I don't think we die easy. — I should like to 'drop' off, gently, but quickly, some 30 or 40 years hence."

After another examination, Ada's doctors revised their diagnosis and informed William. What he was told upset him so much that he rushed to the Royal Leamington Spa, where Lady Byron was taking a treatment. He arrived at eleven o'clock, just as she was going to bed. Lady Byron's reaction to the report of her daughter's dire illness stunned him. Instead of discussing it or asking questions, she unleashed a furious denunciation of Ada's gambling debts and then excoriated William, blaming the potential financial scandal on his own careless supervision of Ada. *Didn't he know that genius was always a child?* she scolded. William retreated in a state of shock.

Ada had cancer of the uterus, but in keeping with the attitudes of the time, she herself was not told; women were to be spared such harsh truths. The doctor assured William that Ada's illness could be managed. He even claimed that her cancer might be cured.

Ada still contended with Lady Byron's high expectations of her. Ada told her she longed to feel that she had not lived in vain. Envisioning metaphorical soldiers who would do her work, she mused, "Of *what materials* my *regiments* are to consist, I do not at present divulge. I have however the hope that they will be most *harmoniously* disciplined troops; — consisting of vast *numbers* & marching in irresistible power to the sound of *Music.* . . . Certainly *my* troops must consist of *numbers,* or they can have no existence at all."

Chapter 17

An Awful Death

By 1852, Ada's pain was unremitting. She endured it with extraordinary courage and something like cheerfulness, writing reassuring letters to her children and even to her mother. Wine and various powders were prescribed along with laudanum. One of her friends told her about cannabis, and she tried it, with good results. She enjoyed a spurt of mental energy, entertaining scientific and mathematical friends who found her intellect intact and were amazed by her fortitude and optimism. Babbage visited her often. William observed Ada and Babbage's "constant philosophical discussions begetting only an increased esteem & mutual liking."

She kept her mother informed of her condition and treatments, but refused to see her. Lady Byron stayed away, even though Ada and William were living in a house she owned, on Great Cumberland Place, in London. She spitefully sent word to William that she would

Near death, Ada took some solace from her piano.

assume responsibility for Ada's debts, except those for gambling. She demanded a minute accounting of every expenditure, however minor. William had borrowed money from her, too, to pay for construction costs at two of their homes. Lady Byron used the loan to humiliate him.

When Lady Byron discovered that Ada had pawned the Lovelace family jewels to pay off bookmakers, she reacted with fury, insisting that Ada stop taking opium for her pain, because pain was delivered by God for a purpose, in this case to redeem Ada's sins: "The greatest of all the mercies shown her has been her disease — weaning her from temptation, & turning her thoughts to higher and better things," she told a like-minded friend.

William would never forgive Lady Byron for her inhumanity, but he seemed to have been helpless to combat it. He protested the ban on opium but nevertheless carried it out. Lady Byron insisted on sending a pair of mesmerists, who finally admitted their treatments were a failure. Ada suffered in agony, unable to sleep and vomiting everything she ate. She remarked that her father had died at exactly her own age. So had his father, "Mad Jack" Byron, an army officer who squandered his wife's fortune and deserted her. Did she not belong to their line?

Doctors told William that Ada would be dead in two months. But she rallied enough to be propped up at the piano and play a little. Sometimes she was able to look over her scientific papers, smiling like a blessed martyr.

John Crosse visited briefly and Ada managed to slip him the jewels that her mother had redeemed so he could pawn them again and settle her debt (but perhaps also to buy Crosse's silence about their relationship). He was certainly a duplicitous character, having passed himself off as a bachelor, though he had a wife and children. After Ada's death, he demanded to be paid to destroy the letters she had written to him.

Babbage came to see Ada on August 12, during which visit she gave him a letter that named him her executor. He told friends afterward that she sensed she was dying and had lost control of her household.

All the Lovelace children were urgently summoned: Byron from the navy, which William had forced him to join at thirteen; Ralph from school; and Anne from the home of family friends, where she had been staying. Ada was in too much pain to see them for more than a few minutes at a time, but they revived her spirits.

By the time Ada's doctor at last told her the truth about her condition, she had long since accepted it. Mercifully, she was given morphine to ease the pain. At times chloroform was also administered, but she hated its effects: "To be dead sleepy, & absolutely *torn* awake by agony, & obliged to use every imaginable resource again & again for hours!" She told William she wanted to be buried next to her father, and he began making the arrangements.

On August 19, 1852, Ada asked to see her friend Charles Dickens. He came quickly and, at her request, read her the death scene from *Dombey and Son,* her favorite of his books. He was the last person outside the family to see her and was struck by her courage.

When Lady Byron burst on the scene, she promptly dismissed Ada's servants, including her personal maid. She declared Ada's letter naming Babbage her executor legally unenforceable and forbade Babbage to visit again. Ada begged to see him, and even her doctor said she mustn't be prevented from receiving her dear friend, but Lady Byron would not be moved. It was again her house.

On September 1, a drugged and agonized Ada was ordered by Lady Byron to save her soul by confessing her misdeeds to William. William went to Ada's room and shut the door. A few minutes later, he emerged, trembling with emotion, his face ashen. He walked away quickly and never told anyone what he'd heard. He merely scrawled a note saying that, should he be absent, "Lady Byron is the Mistress of My House."

It took nearly three more months for Ada to die, and all that time, a righteous Lady Byron kept vigil at her bedside. She commissioned a sermon from Dr. King on the efficacy of suffering to absolve sin and may have read it to her tormented daughter. Ada died on the evening of November 27, 1852.

On the third of December, Ada was buried in the little village church at Newstead Abbey. One of Lady Byron's friends thought the funeral "was too ostentatious — escutcheons and coronets everywhere." Lady Byron did not attend.

In her short, tormented life, Ada Byron Lovelace knew the exhilaration of stretching her great mind to its very limits and visualizing what was still inconceivable to others. With her extraordinary gift for "poetical science," she was the first person to write about what would come to be known as computer programming. With that achievement, Ada Lovelace had launched her imagination a full century into the future.

Babbage had captured Ada's otherworldly power by describing her to Michael Faraday as "that Enchantress who has thrown her magical spell around the most abstract of Sciences and has grasped it with a force which few masculine intellects . . . could have exerted over it."

And so we remember her.

Lady Byron's sketch of Ada on her deathbed

EPILOGUE

Charles Babbage

In 1854, an American professor met Babbage and made notes of their conversation. Babbage talked about Ada's unique gifts as a mathematician. He said that she alone had been capable of describing his machine, but

> *it was the recollection of her miserable life — he spoke of it as a tragedy — that seemed to sadden him. . . . There was so much feeling in both his words and manner that I did not feel at liberty to question him as to the precise nature of the unhappiness of the life he was speaking of and its tragic termination. . . . I gathered that "Ada" had a good deal of the Byron devil in her, and that having made an uncongenial match with Lord Lovelace, she cordially disliked him, and that she also had no better feeling for her own mother; it seems to have been a case of triple antipathy between the wife, and husband, and mother. Speaking*

of Lady Lovelace's matter-of-fact mind, Mr. Babbage told me he used to have a good deal of good-natured fun by telling her all sorts of extraordinary stories.

Ada and Babbage's collaboration on Ada's paper was a brilliant highlight of their lives, a time when both were driven and full of hope and accomplishment.

Models of the never-built Difference and Analytical Engines that Babbage designed are exhibited in the London Science Museum. The Smithsonian Institution in Washington, D.C., owns a version built in Sweden in 1853. In 1989–1991, a full-size Difference Engine was finally built according to his blueprints and is housed in the London Science Museum.

Charles Babbage lived a long life and died in 1871 a bitter man. The future had excited Babbage. He said once that he would gladly give up what time he had left if he could live five centuries into the future for three days. Prophet though he was, he didn't foresee the incredible acceleration of progress that would bring the future he dreamed of much sooner than five hundred years.

Lady Byron

Lady Byron cancelled the bequests Ada had listed in her letter to Babbage. He ended up with a single book. The rest of the items intended for him (and others) went to Lady Byron's friends. Ada's servants were made to sign statements stating they would accept reduced payments that they didn't deserve. Ada's maid Mary refused to sign and got nothing until Babbage left her some money himself.

Charles Babbage, circa 1847–1851

Lady Byron destroyed most of Babbage's letters to Ada. She then tried to get Babbage to turn over the letters Ada had written to him, and he indignantly refused. Throughout her life, she kept copies of her own correspondence wherever it reinforced her versions of events.

Lady Byron refused to make amends with William after Ada died, despite his many attempts to placate her. Woronzow pled on his behalf as well but finally had to remind William in an understatement, "You know she is very peculiar in her views, opinions, and actions. These you cannot expect to change."

Lady Byron nourished her reputation as the victim of Lord Byron's wickedness until she died, while refusing to denounce him for it (except privately). Harriet Beecher Stowe interviewed her in old age and enthusiastically subscribed to the story of Lord Byron's shocking misdeeds and Lady Byron's victimhood. When Stowe related it to American readers, her own reputation was tarnished. By that time, Lord Byron's was on the rise again.

William, Lord Lovelace
William lived to be eighty-eight, depriving his surviving children of their inheritance for all that time. In 1865, he married a pleasant woman who tried to make peace between the children and their father.

Children of Ada and William Lovelace
Byron King-Noel, Viscount Ockham, was the child who most resembled his mother and was probably Ada's favorite. She lost him

in effect when William sent him off to the navy at thirteen. Byron seems to have deserted the navy and was never forgiven for it by his father. He refused to accept his title and became a laborer. The physique it gave him dazzled everyone he met. Like his mother and grandfather, he died young.

Anne Blunt, 15th Baroness Wentworth, grew up adoring her grandmother, Lady Byron, and was estranged from her father. She would be as passionate about horses as her mother was about mathematics. She introduced the Arabian breed to England, was a student of Arabic, and owned and played a Stradivarius violin. She lived to the age of eighty, long separated from her handsome and philandering poet husband.

Ada's youngest child, Ralph King-Milbanke, 2nd Earl of Lovelace, accepted his aristocratic station, ultimately inheriting William's title and what was left of his estate. He was more or less raised by his grandmother, Lady Byron, and was her defender. He collected and edited the family papers and tried to make sense of the stories that had dogged the Byron descendants through the generations.

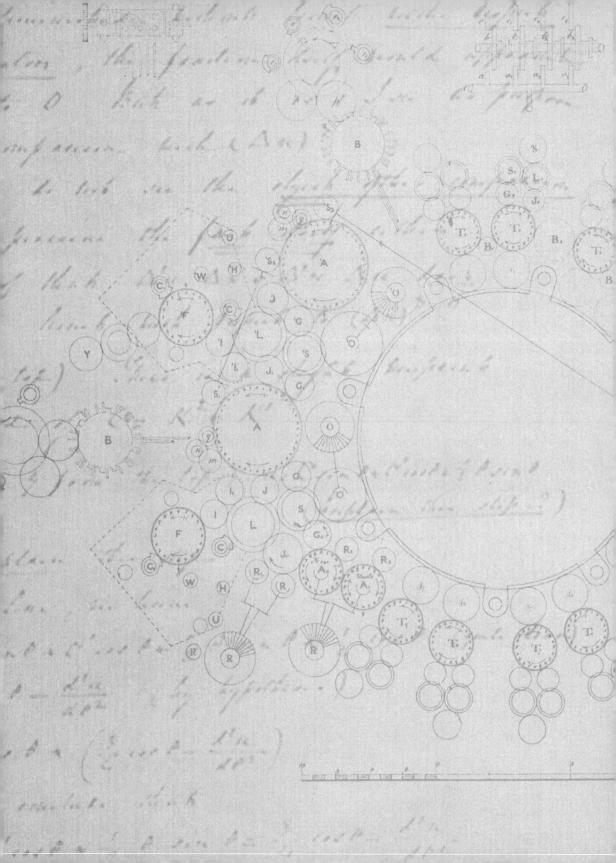

AFTERWORD

Today, computers are very much a part of our lives and we take them for granted, although most of us are unable to explain exactly how these complex machines work. A debate about artificial intelligence, or how much "thinking" computers can do, has raged for decades. Alan Turing, a World War II codebreaker and computer pioneer, coined the term "Lady Lovelace's Objection" after reading Ada Lovelace's warning that the Analytical Engine could only perform programs that people instructed it to perform. In other words, she reasoned that computers are not capable of original or independent thought; they are only as smart as the data and functions input by humans. Thus, the name Lovelace became forever linked to the debate over artificial intelligence. Turing actually disagreed with the objection.

Babbage's models and drawings were placed in a London museum, where some people saw them. Mathematicians like Turing who worked in code-breaking operations at Bletchley Park during World War II spoke among themselves of Babbage's inventions.

The first developers of electronic computers sometimes cited Charles Babbage and Ada Lovelace as inspirations. But Babbage's very cumbersome machine was not built in his lifetime, and Lovelace's essay, neglected for a hundred years, was only republished in 1953 in B. V. Bowden's groundbreaking book, *Faster than Thought: A Symposium on Digital Computing Machines.* Ada's work is more commonly remembered as a premature burst of insight with no obvious or actual use in what was then a mechanical rather than an electronic age. A great many steps lay between Ada's interpretation of the Analytical Engine and the devices we use today for every kind of task.

Lovelace's extraordinary insights have led her to be labeled "the first computer programmer." In 1980, the U.S. government named a computing language Ada in her honor. People passionate about computing celebrate Ada Lovelace Day every October 11. Outstanding contributors to digital science can now earn the Ada Lovelace Award.

APPENDIX A

ADA'S NOTES

Here is a summary of Ada Lovelace's notes in "Sketch of the Analytical Engine Invented by Charles Babbage" by L. F. Menabrea. Menabrea's work was originally published in French in the *Bibliothèque universelle de Genève* 82 (October 1842). Ada's translation, with her extensive notes, was originally published in *Scientific Memoirs* 3 (1843) 666–731. The text is available online at http://www.fourmilab .ch/babbage/sketch.html.

NOTE A

In Note A, Lovelace distinguishes between the Difference Engine and the Analytical Engine. The first can compute values of any polynomial that has a degree not greater than six. The latter can perform operations such as addition and multiplication as well as operations that modern computers can do such as conditional branching and looping. She states in Note A that "the Analytical Engine is an embodying of the science of operations, constructed with peculiar reference to abstract number as the subject of those operations." She also states, "We may consider the engine as the material and mechanical representative of analysis, and that our actual working powers in this department of human study will be enabled more effectually than heretofore to keep pace with our theoretical knowledge of its principles and laws, through the complete control which the engine gives us over the executive manipulation of algebraical and numerical symbols." In other words, she seems to foresee how such a machine would expand our knowledge!

The operations of the Analytical Engine can be "any process which alters the mutual relation of two or more things," including "all subjects in the universe." This versatility is due to the use of punch cards like those in a Jacquard loom; the Analytical Engine "weaves algebraical patterns just as the Jacquard-loom weaves flowers and leaves." One type of card specifies the operation, such as a mathematical operation. Another type of card holds numerical values. The third type of card directs the transfer of values from the Store (or memory) to the Mill (where operations are performed), and of results from the Mill to the Store.

The engine uses a new language for analysis — in other words, what we call code. She also urges the government to support construction of the engine, fearing that some other government would undertake it first and England's reputation and future would be damaged.

NOTE B

In Note B, Lovelace describes the immensity of the machine, which contains more than two hundred columns, each with up to forty independently turning discs or gears making up the Store, or memory. Then, assigning values to variables *a*, *x*, and *n*, she shows the steps in solving a complex problem. The solution, she states, "clearly separates those things which are in reality distinct and independent, and unites those which are mutually dependent." In this note, she also introduces the use of inscriptions — notes that act as a "memorandum for the observer, to remind him of what is going on." These inscriptions are similar to remarks, or REMs, used in modern computing — comments from the programmer that describe what a specific piece of code does.

NOTE C

In Note C, Lovelace details that a set of operations called a cycle could be performed multiple times since at the end of a cycle, cards are returned to their original positions. This process is called backing by Babbage and is known as looping today. It enabled the engine to use far fewer cards than the Jacquard loom did, because the cards could be used repeatedly. Lovelace urges her readers to study the principles of the Jacquard loom by visiting one of two Jacquard looms in operation in London.

NOTE D

In Note D, Lovelace supplies a chart showing the steps that save incremental results for use in multiple calculations. She describes the three tall columns containing data: the first column contains data, the next column contains stored data to be used later, and the third column contains the results. She also states that several sets of simultaneous operations can be performed by the Analytical Engine. And when a solution can also be reached in more than one manner, the engine will find the fastest one.

NOTE E

In Note E, Lovelace solves a complicated problem using cycles, or loops (with cards returning to their original position and being used again). She even makes use of loops within loops. Conditional results like the ones she lays out here work like "if-then" statements in modern computing, directing the engine to take the next steps. She cautions against assuming the engine does arithmetic, explaining that in its use of values as abstractions, it is, rather, doing algebra. Furthermore, it has the capacity to manipulate infinite series to determine functions such as logarithms, sines, and tangents.

NOTE F

In Note F, Lovelace shows how using cycles reduces the number of cards needed to solve a problem. The example she uses is eliminating nine variables from ten equations. She emphasizes the simplicity of the Analytical Engine compared to the Jacquard loom, asserting that three cards in the Analytical Engine can do the work of three hundred "or even thousands or millions of cards" in a Jacquard loom.

NOTE G

In Note G, Lovelace makes her most famous assertion: that the Analytical Engine has "no pretensions whatever to originate anything. It can do whatever we know how to order it to perform. It can follow analysis, but it has no power of anticipating any analytical relations or truths."

In this note, she also presents an algorithm for calculating Bernoulli numbers, which have no apparent pattern, by means of looping and branching. This algorithm is considered the first for a "computer," or the first computer code.

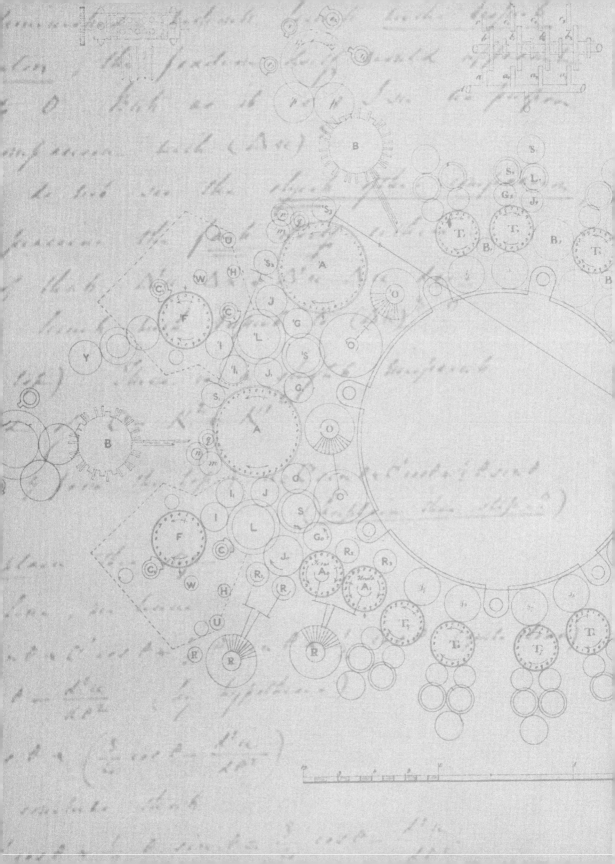

APPENDIX B

THE BRITISH ASSOCIATION FOR THE ADVANCEMENT OF SCIENCE DECLINES TO CONSTRUCT AN ANALYTICAL ENGINE

In August 1878, the British Association for the Advancement of Science met in Dublin and considered supporting the construction of Babbage's Analytical Engine. Here are their thoughts in deciding against doing so:

IX. General Conclusions, and Recommendation

1. We are of opinion that the labours of Mr. Babbage, firstly on his Difference Engine, and secondly on his Analytical Engine, are a marvel of mechanical ingenuity and resource.

2. We entertain no doubt as to the utility of such an engine as was in his contemplation when he undertook the invention of his analytical engine, supposing it to be successfully constructed and maintained in efficiency.

3. We do not consider that the possibilities of its misuse are any serious drawback to its use or value.

4. Apart from the question of its saving labour in operations now possible, we think the existence of such an instrument would place within reach much which, if not actually impossible, has been too close to the limits of human skill and endurance to be practically available.

5. We have come to the conclusion that in the present state of the design of the engine it is not possible for us to form any reasonable estimate of its cost, or of its strength and durability.

6. We are also of opinion that, in the present state of the design, it is not more than a theoretical possibility; that is to say, we do not consider it a certainty that it could be constructed and put together so as to run smoothly and correctly, and to do the work expected of it.

7. We think that there remains much detail to be worked out, and possibly some further invention needed, before the design can be brought into a state in which it would be possible to judge whether it would really so work.

8. We think that a further cost would have to be incurred in order to bring the design to this stage, and that it is just possible that a mechanical failure might cause this expenditure to be lost.

9. While we are unable to frame any exact estimates, we have reason to think that the cost of the engine, after the

drawings are completed, would be expressed in tens of thousands of pounds at least.

10. We think there is even less possibility of forming an opinion as to its strength and durability than as to its feasibility or cost.

11. Having regard to all these considerations, we have come, not without reluctance, to the conclusion, that we cannot advise the British Association to take any steps, either by way of recommendation or otherwise, to procure the construction of Mr. Babbage's Analytical Engine and the printing tables by its means.

12. We think it, however, a question for further consideration whether some specialized modification of the engine might not be worth construction, to serve as a simple multiplying machine, and another modification of it arranged for the calculation of determinants, so as to serve for the solution of simultaneous equations. This, however, inasmuch as it involves a departure from the general idea of the inventor, we regard as lying outside the terms of reference, and therefore perhaps rather for the consideration of Mr. Babbage's representatives than ours. We accordingly confine ourselves to the mere mention of it by way of suggestion.

SOURCE NOTES

INTRODUCTION

p. ix: "discoverer of the *hidden realities* of nature": quoted in Gleick, p. 113.

p. ix: "immense reasoning faculties": quoted in Toole, p. 101.

p. ix: "Fairy": ibid., p. 146.

p. ix: "Enchantress": quoted in Swade, p. 165.

p. xi: "You know I am . . . I think not": quoted in Toole, p. 83.

CHAPTER 1: BORN INTO SCANDAL

p. 3: "When shall we three meet again?" and "In Heaven": Elwin, *Lord Byron's Wife,* p. 409.

p. 4: "There is a pleasure . . . yet cannot all conceal": Byron, *Childe Harold's Pilgrimage, Canto the Fourth,* p. 92.

p. 4: "mad, bad, and dangerous to know": quoted in Mayne, p. 33.

p. 5: "How wonderful of that sensible . . . Crimes & Rivals": quoted in Moore, p. 24.

p. 5: "habitual passion . . . ardent temperaments": quoted in Grosskurth, p. 238.

p. 5: "an Ennui . . . the same origin": ibid., pp. 238–239.

p. 5: "written under a delusive feeling in its favor": ibid., p. 238.

CHAPTER 2: MOTHER AND CHILD WITH GOVERNESSES

p. 8: "Ada has cut two more teeth . . . Want of Mercy": quoted in Elwin, *Lord Byron's Family,* p. 51.

p. 8: "selfish way of assuming . . . the Child's temper": ibid., p. 165.

p. 9: "Wipe away her tears . . . I good" and "Throw me in!": quoted in Moore, p. 18.

p. 9: "When our child's . . . she must forego?": George Gordon Byron, "Fare Thee Well," in Byron, *Complete Works,* p. 876.

p. 9: "I have a great love . . . which I hope not": quoted in Essinger, p. 40.

p. 10: "Is thy face . . . Whither I know not": Byron, *Childe Harold's Pilgrimage, Canto the Third,* p. iii.

p. 10: "My name . . . on my forehead": quoted in Elwin, *Lord Byron's Wife,* p. 450.

p. 12: "rational order of education from the cradle": quoted in Moore, p. 10.

p. 12: "My daughter . . . is rather difficult": quoted in Stein, p. 22.

p. 12: "I have engaged a person . . . educate Ada sufficiently": quoted in Elwin, *Lord Byron's Family,* p. 212.

p. 12: "Be most careful . . . fancies into her head": ibid., p. 168.

p. 12: music, 10 . . . French exercises, 4:30: ibid., p. 229.

p. 13: "At night she cannot go . . . easily worked upon" and "brim full . . . completely happy": ibid., p. 212.

p. 13: Swiss educator and philanthropist: The Swiss educator whom Lady Byron admired was named Emanuel von Fellenberg. Lady Byron had consulted with him when setting up a school for homeless children. (Around 1835, Ada was a volunteer teacher of arithmetic there.) Fellenberg, in turn, was a disciple of Johann Heinrich Pestalozzi, whose educational theories stressing freedom of exploration were influential in Europe and America. Albert Einstein, after running away from a rigid German school, found refuge in a Swiss one designed by Pestalozzi.

Sophia De Morgan's memoirs contain a summary that parallels Lady Byron's theory of education: "Teaching will form character. No matter from what parents a child is born, give him but suitable training, and you may make what you please of him" (quoted in Baum, p. 8).

p. 14: "G" for good and "B" for bad: Elwin, *Lord Byron's Family*, p. 228.

p. 14: "I want to please . . . & never move": ibid., p. 217.

pp. 14–15: "Miss Lamont appears . . . from those she loves": ibid., p. 215.

p. 15: "the *passion* She has taken . . . does not spoil her": ibid., p. 165.

p. 15: "Ada loves me . . . fond of me": quoted in Mayne, pp. 280–281.

CHAPTER 3: SHE HAS A FATHER

p. 17: "fullness of the vessels of the head": quoted in Moore, p. 21.

p. 18: "prevailing characteristic" and "Cheerfulness — a disposition to enjoyment": ibid., p. 22.

p. 18: "account of Ada's disposition . . . fool in a family": quoted in Elwin, *Lord Byron's Family,* p. 236.

p. 20: "if Grandpapa & Papa were the same": quoted in Stein, p. 25.

p. 20: "a feeling of dread . . . the day of her death": ibid.

p. 20: "like an earthquake": Allan Cunningham, "Robert Burns and Lord Byron," *London Magazine,* August 1824, 117–124, http://spenserians .cath.vt.edu/CommentRecord.php?action=GET&cmmtid=7507.

p. 21: "Ada! sole daughter . . . they smiled": Byron, *Childe Harold's Pilgrimage, Canto the Third,* p. iii.

p. 21: "I should wish . . . I had a reward": quoted in Toole, p. 22.

p. 21: "conversational litigation": ibid., p. 30.

CHAPTER 4: HER IMAGINATION SOARS

p. 27: "If 750 men . . . 1200 men require": quoted in Toole, p. 24.

p. 28: "Fairy": ibid., p. 146.

p. 28: "Enchantress": quoted in Swade, p. 165.

p. 29: "bring the art of flying . . . great perfection": quoted in Toole, p. 25.

p. 29: "Carrier Pigeon": ibid., p. 27.

CHAPTER 5: GUARDED BY FURIES

p. 31: "There are no weeds . . . she liked particularly": quoted in Essinger, p. 55.

p. 32: "had *no* taste . . . kind feeling": quoted in Moore, p. 34.

p. 33: "like a fish": ibid., p. 29.

p. 33: "Furies": quoted in Toole, p. 32.

CHAPTER 6: AN ELOPEMENT

p. 34: "had the same love . . . her father had": quoted in Moore, p. 34.

p. 35: "went as far . . . connexion": quoted in Woolley, p. 120.

p. 35: "objectionable thoughts": quoted in Stein, p. 43.

p. 36: "hard work": ibid.

p. 36: "You will soon puzzle me in your studies": ibid., p. 45.

p. 36: "I must cease . . . want of excitement.": quoted in Toole, p. 40.

p. 36: "constituted my guardian by God *forever*": quoted in Toole, p. 34.

p. 37: "calmly": quoted in Mayne, p. 323.

p. 37: "I trust to the principles . . . *dare* to be poor": ibid.

p. 37: "I would rather . . . in the *wrong*": ibid., p. 324.

CHAPTER 7: MEETING BABBAGE

p. 41: a wide range of reckless behaviors: Lady Byron was related to aristocrats who misbehaved. Her aunt was Lady Melbourne, Lord Byron's close confidante and herself the object of gossip after she bore children not by her husband. Lady Melbourne encouraged Byron's love affairs as well as his marriage. Her own daughter-in-law Caroline Lamb amused society when she openly pursued Byron. It was Caroline who called Byron "mad, bad, and dangerous to know."

p. 42: "than any assemblage in the *grand monde*": quoted in Essinger, pp. 80–81.

p. 44: "Difference Engine": See Swade for a discussion of this machine.

p. 45: To produce accurate numerical tables . . . : For further discussion of the calculation of differences, see Babbage, *Charles Babbage*, p. 227.

p. 47: "miracles": quoted in Swade, p. 76.

p. 47: "While other visitors . . . beauty of the invention": ibid., p. 167.

p. 49: "the whim of the moment": quoted in Toole, p. 39.

CHAPTER 8: A ROLE MODEL

p. 50: "intensely ambitious . . . my early days": quoted in Holmes.

p. 53: "overwhelming incumbrance of numerical detail": quoted in Swade, p. 129.

p. 53: "throwing a bridge . . . unknown" and "It occurred . . . to foresee": quoted in Toole, p. 52.

p. 53: "unsound and paradoxical" and "universality": ibid.

p. 54: "gem of all mechanism": ibid., p. 46

p. 54: "holes in a set . . . by the artist": Babbage, *Charles Babbage*, p. 55.

CHAPTER 9: COURTSHIP AND MARRIAGE

p. 55: "made a very daring . . . an heiress": quoted in Stein, p. 52.

p. 55: "elopement": ibid., p. 36.

p. 55: "imperfection of [her] actions" and "far from what they should be": Ada Byron Lovelace to Mrs. King, April 12, 1834, Carl H. Pforzheimer Collection of Shelley and His Circle, New York Public Library.

p. 55: "ordinary music" and "exciting tendency of . . . might lead to": ibid.

p. 56: "Since we parted . . . than in a dream": quoted in Moore, p. 65.

p. 56: "an unexpected happiness" and "calm": quoted in Toole, p. 56.

p. 57: "enviable calmness and philosophy": quoted in Moore, p. 67.

p. 57: "has escaped . . . morn of her life": quoted in Baum, p. xviii.

CHAPTER 10: MOTHERHOOD

p. 61: "So you see . . . for those pursuits": quoted in Toole, p. 63.

p. 61: "prove the Mrs. Somerville of Housekeepers!" and "produce a Treatise . . . Domestic Sciences'": quoted in Stein, p. 55.

p. 63: "I have made . . . of my Henship!": quoted in Moore, p. 74.

p. 63: "Dearest Son": ibid., p. 152.

p. 64: "little treasure" and "leave me in peace": quoted in Toole, p. 69.

p. 64: "just the person . . . fair way for it": quoted in Stein, p. 58.

CHAPTER 11: GRASPING FOR MATHEMATICS

p. 67: "I have a peculiar *way* . . . teach me successfully": quoted in Toole, p. 82.

p. 67: "You know I . . . I think not": ibid., p. 83.

p. 67: "the greatest favour any one can do me": ibid.

p. 69: "the Algebra . . . In Differentiation": quoted in Toole, p. 86.

p. 69: "I have more . . . of *your* business": ibid., p. 104.

p. 69: "The Professor . . . one I imagine": ibid., p. 88.

p. 69: nervous ailments that had plagued Ada most of her life: Regarding Ada's health and its impact on her studies, see Stein, pp. 80–83.

p. 70: "power," "an original mathematical investigator, perhaps of first-rate eminence," and "prophesied": quoted in Stein, p. 82.

p. 70: "would but attend to . . . imposed on it": ibid., pp. 81–82.

p. 70: "things hidden . . . ordinary senses" and "immense reasoning . . . *one* vast focus": quoted in Toole, p. 101.

p. 71: "The resemblance . . . a little daft": quoted in Moore, p. 160.

CHAPTER 12: THE NEW ENGINE

p. 72: "I am myself . . . have given it": quoted in Swade, p. 118.

p. 72: "*My head* may . . . purposes & plans": quoted in Toole, p. 97.

p. 75: the dream of automation: Imogen Forbes-MacPhail brilliantly discusses automation in the Romantic period in her lecture "The Analytical Engine and the Aeolian Harp" delivered at Oxford University's Ada Lovelace Symposium, December 18, 2015, 21:41, http://podcasts.ox.ac.uk/analytical-engine-and-aeolian-harp.

p. 76: "thinking machine": quoted in Toole, p. 38.

p. 77: Lady Byron finally revealed . . . : On Lady Byron's revelation and Ada's reaction, see Mayne, pp. 354–356.

p. 77: "how you ever came . . . decidedly was not": quoted in Toole, p. 112. William never believed that Medora was Byron's child. Perhaps Ada's saying the incest could occur only to a "very depraved and vicious mind" suggests that she had her doubts, as well.

p. 77: "If my poor . . . philosophical turn!": ibid., p. 110.

p. 77: "an ambition to . . . genius," "less flash," "more depth," and "bring out great . . . task to me!": ibid., p. 112.

CHAPTER 13: A RESTLESS STUDENT

pp. 78–79: "nothing less than . . . laws are known": quoted in Babbage, *Charles Babbage*, p. 226.

p. 79: "algebraical notation": ibid.

p. 79: "eating its own tail": quoted in Gleick, p. 118.

p. 79: "Mathematical Scrap-Book" and "main points . . . for desiring this": quoted in Toole, p. 93.

p. 80: "First: it is the *Combining* . . . Discovering Faculty": ibid., p. 94.

p. 80: "Mathematical Science . . . the unconscious": ibid.

p. 80: "add my mite . . . knowledge of the world": ibid., p. 96.

p. 81: "bond of union . . . & *muscular action*": quoted in Stein, p. 132.

p. 81: "unnatural feelings mental & bodily": ibid., p. 81.

p. 81: "all my ill . . . foundations in that": quoted in Toole, p. 205.

CHAPTER 14: MASTERWORK

p. 85: "gigantic idea": quoted in Babbage, *Charles Babbage*, p. 226.

p. 86: "Be kind enough . . . for me": quoted in Toole, p. 141.

p. 86: "You were a little . . . & inaccurate": ibid., p. 154.

p. 86: "I wish . . . am myself": ibid., p. 157.

p. 87: "My Dear Babbage . . . unavoidable engagements": ibid., p. 144.

p. 87: "I am doggedly . . . very bottom" and "I am grappling . . . it with others": ibid., p. 147.

p. 87: "the fundamental relations . . . science of harmony" and "the engine might . . . complexity or extent": Menabrea, Note A.

pp. 87–88: "Many persons . . . other *general* symbols": ibid.

p. 88: "All this was . . . the noblest metal": quoted in Toole, p. 141.

pp. 88–89: "We discussed together . . . in the process": Babbage, *Charles Babbage,* p. 68.

p. 90: "Before ten years . . . system of mine": quoted in Toole, p. 147.

p. 90: "quagmire . . . these *Numbers*" and "Yours puzzle-pate": ibid., p. 149.

p. 90: "I have worked . . . scrupulously attended to": ibid., p. 143.

pp. 90–91: "It is desirable . . . acquainted with": ibid., pp. 191–192.

p. 91: "science of operations": See Toole, chapter 12, for a discussion of this topic.

p. 92: "This science constitutes . . . with each other": Menabrea, Note A.

p. 94: "What shall we . . . his calculating machine?": quoted in Swade, p. 136.

p. 94: "suicidal": quoted in Toole, p. 160.

p. 94: "I am sorry . . . to do with" and "He will never forgive me": ibid., p. 161.

p. 95: "Out of *several* . . . for such negligence": ibid., p. 156.

p. 95: "I do not think . . . just alike)": ibid.

p. 95: "I do *not* believe . . . (& Metaphysician)": ibid., 156–157.

p. 95: "The *pithy. . .* of the writing": ibid., p. 155.

p. 95: "It is not my wish . . . A. A. L.": ibid., p. 145.

pp. 95–96: "The lady who we know . . . in the literary world": Richard Holmes, "Will You Concede Me Poetical Science?," Ada Lovelace Symposium, Mathematical Institute, Oxford University, December 9–10, 2015, posted December 18, 2015, 18:39–18:56, http://podcasts.ox.ac.uk/will-you-concede-me-poetical-science.

p. 96: "savante" and "Lady Lovely": ibid.

p. 96: "No one can . . . in this country": quoted in Toole, pp. 170–171.

p. 96: "formidable reviewer": quoted in Swade, p. 162.

p. 96: "bring his engine . . . W's suggestion": quoted in Toole, p. 170.

pp. 96–97: "Far be it from *me* . . . it than myself": ibid., p. 168.

p. 97: "What a General you would make!": quoted in Moore, p. 158.

p. 97: "refused all the conditions.": quoted in Essinger, p. 192.

CHAPTER 15: AT LOOSE ENDS

p. 98: "papers enough . . . much admired Interpretess": quoted in Toole, pp. 171–172.

p. 99: "must be struck . . . & wondered at": quoted in Moore, p. 159.

p. 100: "Oh! At least you have lost that *mad* look": ibid., p. 212.

CHAPTER 16: RESTLESS SPIRIT

p. 101: "Galvanization . . . things for me": quoted in Toole, p. 211.

p. 101: *"experimental laboratory* . . . inseparable from me": ibid.

p. 101: "A Newton . . . a crying want": ibid., p. 212.

p. 101: "I hope to . . . *Calculus of the Nervous System*": ibid., p. 215.

p. 102: "Faraday expresses himself . . . of my *intellect*'" and *"rising star of Science"*: ibid., p. 211.

p. 102: "thunder and lightning man": "Andrew Crosse, the Man and the Myth," https://www.nationaltrust.org.uk/fyne-court/features /andrew-crosse-the-man-and-the-myth-.

p. 102: "the bride of science": ibid., p. 215.

p. 103: "the most *unorganized* domestic system I ever saw": ibid., p. 217.

p. 103: "the most *utter* lack . . . chaos & chance": ibid., p. 218.

p. 103: "My character . . . I was 26": ibid., p. 261.

p. 103: *"life-less life"*: ibid., pp. 229–230.

p. 103: "*one continuous . . .* long been so," "It is not his fault . . . certain *social position,*" and "irksome *duties*": ibid., p. 230.

p. 104: "it would do irreparable . . . ways than one": ibid., p. 252.

p. 105: "You will not concede . . . *poetical science?*": ibid., p. 235.

p. 105: "spasms of the heart": quoted in Moore, p. 240.

p. 105: "poor thing" and "she is looking . . . to touch upon": ibid., p. 241.

p. 106: "I want you to *complete* something": quoted in Toole, p. 248.

p. 106: "I hope Babbage . . . age & exterior": quoted in Stein, p. 196.

pp. 106–107: "All is like *death . . .* stone behind it?": quoted in Toole, p. 265.

p. 107: "We ought to . . . winds of Heaven": ibid.

p. 107: "a perfect blank": quoted in Moore, p. 276.

p. 108: "never before met . . . cultivated a lady": ibid.

p. 108: "had a *resurrection*": quoted in Toole, p. 270.

p. 108: "Mythic idea": quoted in Moore, p. 271.

p. 108: "better for them not to have known me": ibid.

p. 108: "best friend": quoted in Mayne, p. 388.

p. 110: "the *local* condition . . . a *healthy* sore": quoted in Toole, p. 284.

p. 110: "dulled": ibid., p. 287.

p. 110: "that horrible *struggle* . . . 40 years hence": ibid., p. 290.

p. 111: "Of *what materials* . . . existence at all": ibid., pp. 291–292.

CHAPTER 17: AN AWFUL DEATH

p. 112: "constant philosophical discussions . . . & mutual liking": quoted in Moore, p. 299.

p. 114: "The greatest of all . . . and better things": ibid., p. 312.

p. 115: "To be dead sleepy . . . again for hours!": quoted in Stein, p. 225.

p. 116: "Lady Byron is the Mistress of My House": quoted in Toole, p. 306.

p. 116: "was too ostentatious . . . coronets everywhere": quoted in Moore, p. 333.

p. 116: "poetical science" : quoted in Toole, p. 235.

p. 117: "that Enchantress who . . . exerted over it": quoted in Essinger, p. xiv.

EPILOGUE

pp. 119–120: "it was the recollection . . . extraordinary stories": Henry Reed quoted in Bledsoe and Herrick, p. 423.

p. 122: "You know she . . . expect to change": quoted in Stein, p. 252.

GLOSSARY

animal magnetism: an alleged force that animates all living things and was thought to be able to be harnessed for healing purposes

Bernoulli numbers: a sequence of rational numbers that appears often in mathematics. A rational number is one that can be divided by two integers (whole numbers rather than fractions). The first four Bernoulli numbers are 1, ±1/2, 1/6, and −1/30.

biquadratic equation (or biquadratic function): an equation or function in the form of $ax^4 + bx^2 + c = 0$

capitalist economy: system of goods, services, and exchange in which the means of producing goods and services are owned by private individuals or businesses and in which government plays only a limited role

cubic equation (or cubic function): an equation or function in the form of $ax^3 + bx^2 + cx + d = 0$

cooperative movement: a movement that began in the United Kingdom in the nineteenth century based on the idea that groups of people should own and control industries for their mutual economic benefit

divining rod: a Y- or L-shaped rod that is used to try to locate water, metals, gemstones, or other valuable products; the use of divining rods is not supported by scientific evidence

electrobiology: In the nineteenth century, this term referred to the notion that an electric current could be used to control a subject by stimulating electrons in the subject's body.

electric generator: a machine that converts mechanical energy to electrical energy

electric transformer: a device that transfers electrical energy from one electric circuit to another by means of electromagnetic induction; used in the transmission of electricity from, for example, a power station to homes

electromagnetic induction: the generation of electric current by inducing a change in a magnetic field

electromagnetism: the study electricity and magnetism

function (mathematical): an expression that defines a relationship between one variable and another variable that is dependent on the first one; a mathematical tool that relates an input with one output

galvanization (electrotherapy): the use of electrical energy in medical treatment

gastritis: inflammation of the stomach's mucus layers; can be caused by intake of alcohol or other drugs, food poisoning, or infectious diseases; some symptoms are pain, nausea, loss of appetite, vomiting, and diarrhea

Great Reform Act (1832): an act of the United Kingdom's Parliament that introduced wide-ranging changes to the electoral system of England and Wales

hysteria: a diagnosis that was applied specifically to women and is no longer used by medical professionals; symptoms were said to include anxiety, shortness of breath, fainting, insomnia, irritability, and nervousness, as well as sexually forward behavior. Colloquially, the term is used today to mean an excess of uncontrolled emotion.

Industrial Revolution: the period during the eighteenth and nineteenth centuries in which the emphasis of manufacture and the economy shifted from rural communities centered around agriculture and handcrafts to urban communities centered around industry and factories

logarithm: a shortcut in expressing an exponent $y = a^x$ (in which $x > 0$, $a > 0$, and $a \neq 1$); the corresponding logarithmic equation to exponential equation $y = a^x$ is $\log_a(y) = x$ or the logarithm of y to base a; logarithms can be used to reduce the differences between a group of wide-ranging quantities such as the pH scale for measuring acidity and the scale for the loudness of sound

Luddites: English workers who protested manufacturers who used machines to evade labor laws and lay off workers in the nineteenth century

mesmerism: hypnosis using the philosophy of animal magnetism

method of finite differences: a method of determining a polynomial by using values of the polynomial at certain points

Occam's razor: This principle states that entities should not be multiplied unnecessarily. It is often interpreted to mean that if there are two competing theories that make exactly the same predictions, the simpler theory is the better one.

ophthalmoscope: an instrument that enables one to see the interior of the eye

polynomial: an expression that consists of variables and coefficients; involves only the operations of addition, subtraction, and multiplication; and has only exponents that are non-negative integers

shorthand: a fast method of writing by using abbreviations and symbols; it is useful for taking notes during a speech

socioeconomic status: a descriptor defined by group members' education, income, and occupation; three socioeconomic levels are usually designated (high, middle, and low)

speedometer: an instrument that measures the instantaneous speed of a vehicle

suffrage: the right to vote in political elections

telegraph: an apparatus for transmitting (sending and receiving) messages; an electric telegraph is a machine that transmits coded messages long-distance using electric signals

BIBLIOGRAPHY

"Ada Lovelace Symposium." Mathematical Institute, Oxford University. December 9–10, 2015. Oxford University Podcasts. http://livestream.com/oxuni/lovelace.

"Andrew Crosse, the Man and the Myth." https://www.nationaltrust.org.uk/fyne-court/features/andrew-crosse-the-man-and-the-myth-.

Babbage, Charles. *Charles Babbage and His Calculating Engines.* Edited by Philip Morrison and Emily Morrison. New York: Dover, 1961.

———. *Passages from the Life of a Philosopher.* Edited by Martin Campbell-Kelly. New Brunswick, NJ: Rutgers University Press, 1994.

"The Babbage Engine." Computer History Museum. http://www.computerhistory.org/babbage.

Baum, Joan. *The Calculating Passion of Ada Byron.* Hamden, CT: Archon, 1986.

Bledsoe, Albert Taylor, and Sophia M'Ilvaine Bledsoe Herrick, eds. "Life, Character, and Works of Henry Reed." *Southern Review,* vol. 1 (April 1867).

Byron, George Gordon. *Childe Harold's Pilgrimage.* London: Murray, 1812.

———. *Childe Harold's Pilgrimage, Canto the Fourth.* London: Murray, 1818.

———. *Childe Harold's Pilgrimage, Canto the Third.* London: Murray, 1816.

———. *The Complete Works of Lord Byron.* Paris: A. and W. Galignani and Co., 1835.

Crane, David. *The Kindness of Sisters.* New York: Knopf, 2002.

Elwin, Malcolm. *Lord Byron's Family: Annabella, Ada, and Augusta 1816–1824.* Edited by Peter Thomson. London: J. Murray, 1975.

———. *Lord Byron's Wife.* New York: Harcourt, Brace, 1962.

Essinger, James. *Ada's Algorithm: How Lord Byron's Daughter Ada Lovelace Launched the Digital Age.* Brooklyn: Melville House, 2013.

Gleick, James. *The Information: A History, a Theory, a Flood.* New York: Vintage, 2011.

Graham-Cumming, John. "The Greatest Machine That Never Was." TED Ed, YouTube. 12:14. Posted June 19, 2013. https://www.youtube.com/watch?v=FlfChYGv3Z4.

Grosskurth, Phyllis. *Byron: The Flawed Angel.* Boston: Houghton Mifflin, 1997.

Haigh, Thomas, and Mark Priestley. "Innovators Assemble: Ada Lovelace, Walter Isaacson, and the Superheroes of Computing." *Viewpoints, Communications of the ACM* 58, no. 9 (September 2015): 20–27.

Holmes, Richard. "The Royal Society's Lost Women Scientists." *Guardian,* November 20, 2010. https://www.theguardian.com/science/2010/nov/21/royal-society-lost-women-scientists.

Hyman, Anthony, *Charles Babbage: Pioneer of the Computer.* Princeton, NJ: Princeton University Press, 1985.

Isaacson, Walter. *The Innovators: How a Group of Hackers, Geniuses, and Geeks Created the Digital Revolution.* New York: Simon & Schuster, 2014.

Mayne, Ethel Colburn. *The Life and Letters of Anne Isabella Lady Noel Byron.* New York: Scribner's, 1929.

Menabrea, L. F. "Sketch of the Analytical Engine Invented by Charles Babbage (with Notes Upon the Memoir by the Translator Ada Augusta, Countess of Lovelace)." *Bibliothèque universelle de Genève* 82 (October 1842). http://www.fourmilab.ch/babbage/sketch.html.

Moore, Doris Langley. *Ada, Countess of Lovelace: Byron's Legitimate Daughter.* New York: Harper & Row, 1977.

Morais, Betsy. "Ada Lovelace, the First Tech Visionary." *New Yorker,* October 15, 2013. http://www.newyorker.com/tech/elements /ada-lovelace-the-first-tech-visionary.

Morgan, Sydney. *Lady Morgan's Memoirs: Autobiography, Diaries and Correspondence.* Leipzig, Germany: Bernhard Tauchnitz, 1863.

Murray, Venetia. *An Elegant Madness: High Society in Regency England.* New York: Penguin, 2000.

O'Connor, J. J., and E. F. Robertson. "Augusta Ada King, Countess of Lovelace." MacTutor History of Mathematics Archive, School of Mathematics and Statistics, University of St. Andrews, Scotland. http://www-history.mcs.st-and.ac.uk/Biographies/Lovelace.html.

Padua, Sydney. *The Thrilling Adventures of Lovelace and Babbage: The (Mostly) True Story of the First Computer.* New York: Pantheon, 2015.

"Report of the Committee, Consisting of Professor Cayley, Dr. Farr, Mr. J. W. L. Glaisher, Dr. Pole, Professor Fuller, Professor A. B. W. Kennedy, Professor Clifford, and Mr. C. W. Merrifield, Appointed to Consider the Advisability and to Estimate the Expense of Constructing Mr. Babbage's Analytical Machine, and of Printing Tables by Its Means. Drawn up by Mr. Merrifield. 1878." In *Report of the Forty-Eighth Meeting of the British Association for the Advancement of Science; Held at Dublin in August 1878.* http://www.fourmilab.ch/babbage/baas.html.

Stein, Dorothy. *Ada: A Life and a Legacy.* Cambridge, MA: MIT Press, 1985.

Swade, Doron. *The Difference Engine: Charles Babbage and the Quest to Build the First Computer.* New York: Viking, 2001.

Toole, Betty Alexandra. *Ada, the Enchantress of Numbers: Prophet of the Computer Age.* Mill Valley, CA: Strawberry Press, 1998.

Walker, John, ed. "The Analytical Engine." Fourmilab Switzerland website. http://www.fourmilab.ch/babbage/.

Wolfram, Stephen. "Untangling the Tale of Ada Lovelace." *Stephen Wolfram* (blog). December 10, 2015. http://blog.stephenwolfram.com/2015/12/untangling-the-tale-of-ada-lovelace/.

Woolley, Benjamin. *The Bride of Science: Romance, Reason, and Byron's Daughter.* London: Pan, 2000.

IMAGE CREDITS

INDEX